THE ALUMINIUM MULTINATIONALS AND THE BAUXITE CARTEL

MACMILLAN INTERNATIONAL POLITICAL ECONOMY SERIES

General Editor: Timothy M. Shaw, Professor of Political Science and Executive Director, Later Pearson Institute for International Development, Dalhousie University, Nova Scotia

The global political economy is in a profound crisis at the levels of both production and policy. This series provides overviews and case studies of states and sectors, classes and companies, in the new international division of labour. These embrace political economy as both focus and mode of analysis; they advance radical scholarship and scenarios.

The series treats policy-economy dialectics at global, regional and national levels and examines novel contradictions and coalitions between and within each. There is a special emphasis on national bourgeoisies and capitalisms, on newly industrial or influential countries, and on novel strategies and technologies. The concentration throughout is on uneven patterns of power and production, authority and distribution, hegemony and reaction. Attention will be paid to redefinitions of class and security, basic needs and self-reliance and the range of critical analyses will include gender, population, resources, environment, militarisation, food and finance. This series constitutes a timely and distinctive response to the continuing intellectual and existential world crisis.

Robert Boardman
PESTICIDES IN WORLD AGRICULTURE

Jerker Carlsson and Timothy M. Shaw (*editors*)
NEWLY INDUSTRIALIZING COUNTRIES AND THE POLITICAL ECONOMY OF SOUTH–SOUTH RELATIONS

Steven Kendall Holloway
THE ALUMINIUM MULTINATIONALS AND THE BAUXITE CARTEL

James H. Mittelman
OUT FROM UNDERDEVELOPMENT

John Ravenhill (*editor*)
AFRICA IN ECONOMIC CRISIS

Roger Southall (*editor*)
LABOUR AND UNIONS IN ASIA AND AFRICA

Series Standing Order

If you would like to receive future titles in this series as they are published, you can make use of our standing order facility. To place a standing order please contact your bookseller or, in case of difficulty, write to us at the address below with your name and address and the name of the series. Please state with which title you wish to begin your standing order. (If you live outside the UK we may not have the rights for your area, in which case we will forward your order to the publisher concerned.)

Standing Order Service, Macmillan Distribution Ltd,
Houndmills, Basingstoke, Hampshire, RG21 2XS, England.

The Aluminium Multinationals and the Bauxite Cartel

Steven Kendall Holloway

Associate Professor
St Francis Xavier University, Nova Scotia

M
MACMILLAN
PRESS

First published 1988

Published by
THE MACMILLAN PRESS LTD
Houndmills, Basingstoke, Hampshire RG21 2XS
and London
Companies and representatives
throughout the world

Filmsetting by Vantage Photosetting Co. Ltd
Eastleigh and London

Printed in Hong Kong

British Library Cataloguing in Publication Data
Holloway, Steven Kendall
The aluminium multinationals and the bauxite cartel.—(Macmillan international political economy series)
1. Aluminium industry and trade—History
2. Bauxite—History 3. Trusts, Industrial —History
I. Title
338.8'7 HD9539.A62
ISBN 0–333–42814–5

Contents

List of Tables

List of Figures

1 Introduction

The oil embargo of 1973–4, more than any event since the Second World War, stimulated a great deal of interest in international raw material and commodity cartels. Two major lines of scholarly activity have reflected the diversity of this interest. The first takes the part of the developed states and attempts to assess just how vulnerable the Advanced Industrial States (AIS) are on 'outside' sources of strategic goods. The emphasis from this perspective is on a simple inventory of who has how much of what commodity and what the particular AIS's foreign policy should be doing about it.[1] The second line has examined the problem of cartel formation from the producer country perspective and, where the producers are developing countries, the prospects for New International Economic Order (NIEO). The emphasis here is on where, when, and how cartel power is likely to be attained.[2] Indeed, the success of OPEC stimulated the formation of a series of imitator cartel attempts in bauxite, bananas, iron ore and mercury.

A major problem with nearly all this political literature in both viewpoints is that it focuses on government to government interactions and ignores the role of the institution which is actually extracting, processing and distributing the goods – the multinational corporation (MNC). The implicit assumption of the first line is to view the MNC as a passive, apolitical conduit. The assumption of the second perspective is to view the MNC as an agent of the parent national government, or the link point in a chain of dependence. Sampson (1975) stood alone for sometime as providing an attempt to bring to MNCs into the picture as independent actors albeit in a journalistic, historical manner. Pindyck (1977) is one of the few writers studying cartels who admits that the companies may have some impact on cartel formation. While excluding them from his study of the bauxite cartel, he admits that his model's 'main shortcoming is that it ignores the important regional characteristics of the bauxite, as well as the monopsony power of some of the multinational companies that purchase bauxite'.[3]

In the years immediately following the oil embargo, various investigations of the US Congress began to suggest that the international oil companies may have violated what they considered to be aspects of the US national interest. Since then the attack on these companies has covered a wide range of allegations. At a minimum is the sentiment that the companies gave in too quickly, that they did not have the incentives to challenge OPEC. The following quotation typifies this view:

> By going along with OPEC the companies are guaranteed access to OPEC oil at preferred prices. Guaranteed access means that the major companies will have at least some oil to distribute when other companies do not . . . Under these conditions the companies have no incentive to bargain down or challenge OPEC prices and power.[4]

The extreme position has been to allege that OPEC would not have been possible without the collusion of the oil companies. This is due to the alleged role of the companies in acting as distributors and rationers. Consider this quote from Wyant (1977):

> As the Senate Antitrust and Monopoly subcommittee majority staff puts it, 'OPEC, like any cartel, must perform two basic functions. First, it must agree on the price level it wants. Second, and far more difficult, it must agree on how the members will apportion the necessary production cutback among themselves.' However, OPEC has never been able to agree on a formula for prorationing productions. Instead, OPEC relies on the companies to decide how much should be produced in each member country. Without the companies to prorate production, the cartel could not hold together.[5]

The most commonly cited evidence that the oil companies have benefited from this 'co-operative' arrangement is the profits that the firms made during the early years of the price increases. As vertically integrated firms, the majors were supposedly able to raise the final price well above their increased costs by acting in concert following each crude increase, given an inelastic market of short supply. Graphs such as Figure 1.1 appeared in popular news journals to emphasise this allegation.

Two major problems with this type of analysis are apparent. First, there has been an absence of controls for intervening variables that might explain increased profits and second, there has been an absence of generalisation beyond the oil example to other cases where the extracting firms are likely to benefit from a cartel. These two failings are nearly always attributed to the purported exceptional nature of oil, a proposition that will not be debated here. Consequently, this study focuses on the bauxite cartel (the International Bauxite Association) and the question of benefits that might have accrued to the aluminium MNCs involved in the production of bauxite. Bauxite is a mineral which more closely resembles other metal minerals, and hence avoids the problem of the exceptional nature of oil.

Figure 1.1 Oil profits

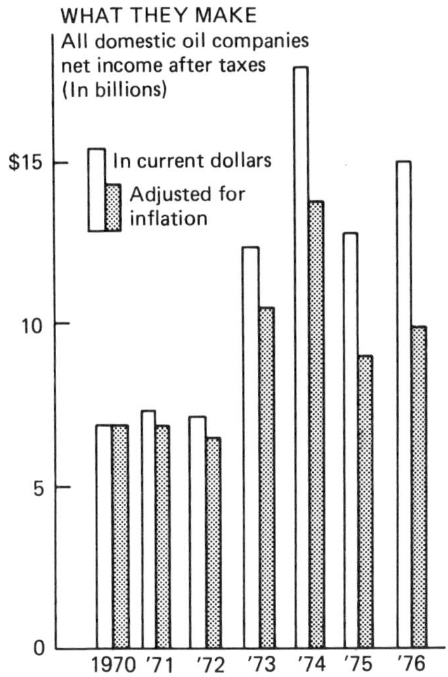

Source: Reprinted with permission from *Time*, 24 October 1977

This study of the bauxite/aluminium industry does not fit neatly into either literature mentioned above. I am concerned here with specifying the conditions under which cartel power may be used by Third World producers but I focus on a feature insufficiently analysed in previous studies. Specifically, this study will suggest a means to analyse the crucial role that MNCs play in producer cartels. I will argue that having at least the acquiescence of the MNCs is a prerequisite to commodity cartel success. The reaction of the companies to the cartel will in turn depend mainly on its impact of their profits and that depends, as I will show, on the underlying characteristics of the industry and markets for the commodity. I will use the aluminium industry and the bauxite cartel as my main referent showing the impact of the cartel on the aluminium multinationals' profits and their subsequent interactions with the bauxite-producing nations.

DEFINITIONS OF CARTEL

The preceding quote from Wyant suggests two important attributes of the formal definition for a cartel which an economist might propose: agreement on price level (assumed to be above the competitive, marginal cost price) and allocation of production (or alternatively, division of markets) among the producers. Caves (1982) goes further and defines a 'contractually complete cartel' in even more restrictive terms.

> It is not enough to divide markets or agree on a common price . . . Without pooling profits and dividing them according to some pre-agreed formula, the members will find that the allocation of production among themselves must meet an inconsistent set of objectives . . . Only with profit-pooling or side-payments can parties attain the maximum joint profit and divide the spoils in a way consistent with their bargaining power.[6]

Formal definitions of this sort are heavily imbued with game-theory calculations and seem to require success in the long term before the label cartel can be applied. Some scholars have even suggested that OPEC itself is not a 'cartel' due to its failure to meet such specific requirements. But Caves himself admits many less successful and less long-term cartel arrangements may be of great benefit to their members. 'It could be rational to enter a cartel expected to be temporary, even if the profits of the cartel members in the competitive period after the cartel's collapse are expected to be less than the competitive profits they earned before its formation' (p. 235). For Caves the defining characteristic of a cartel appears to be the ability to expropriate consumers' surplus (i.e., make monopoly profits) if even for a relatively short period.

From the producing government point of view, cartel formation is a question of raising revenues earned on the nation's primary resource exports. With the adoption of the agenda for the New International Economic Order (NIEO), most Third World states indicated their disillusionment with relying solely on market/forces to establish commodity prices. Indeed, the liberal international economic order was seen as directly contributing to the impoverishment of the less developed countries through declining terms-of-trade and exploitively low prices. The response has been a variety of 'interventionist' moves into the commodity markets including negotiation of international commodity agreements including the consumers and attempts at producer cartels which exclude consumers. From the Third World perspective, 'cartel' implies an attempt by producers to raise commodity prices viewed as historically unfair for the purpose of increasing producers' revenues. Expropriation of consumers' surplus where consumers are mainly wealthly First World nations is seen as a justified and long overdue redistribution of international wealth.

In this study, then, 'cartel' will be applied to any organisation of producers or sellers which internationally and successfully (over at least the 'short-run' period of 2–3 years) manages through collective action to increase sales revenues above the recent rate of return for that commodity. Hence, the word is being used here in a more general and political sense than the formal economic definitions listed above.

This usage does not require a 'formal contract' and may not in the short run require division of markets by cartel members. Wyant suggests that OPEC at least initially relied on the large oil multinationals to regulate the markets using their traditional sources of market power. This arrangement is even clearer in the bauxite/aluminium industry where until recently open markets for bauxite or aluminium did not exist and the industry is dominated by even fewer firms than the case of oil. In such an industry, raising revenues for members may be accomplished simply by collectively raising the price of the commodity and then relying on the business oligopoly to pass on (or administer) the price increase to final consumers. Vernon (1977) refers to MNCs in this position as carrying out a 'tax collector' role for the producers. This study will develop the theme that this in fact has happened at certain points for bauxite, uranium and perhaps oil.

One other point of usage should be established at this time. 'Private' or 'business cartel' will be used to designate a cartel where all members are MNC companies. An example of a business cartel used in this study would be the Aluminium Alliance described in Chapter 3. This organisation clearly fulfills the requirements of even more formal definitions of cartel as it possessed explicit mechanisms for setting price and for dividing production and markets among members. By contrast, a 'governmental producers' cartel' or 'governmental cartel' refers to a cartel as described above in which the membership is predominately national governments and the cartel succeeds in increasing the revenues earned by members on commodities produced or sourced in their territories. An example of a governmental cartel would be OPEC.

Most of the cartels discussed here will be of a mixed mode – governmental cartels which involve at least some co-operation from the private sector. The clearest example of this type is the uranium cartel. The founding of that cartel in Paris in February 1972 involved two sets of simultaneous meetings: governmental and business. In fact, the meetings overlapped directly for actors such as Canada which had governmental representatives at the former and Eldorado (a government-owned company) at the latter.

In conclusion, it must be said that any definition of cartel is likely to be incomplete due to the nature of the subject matter. 'Imperfect competition' as formally defined by economists (i.e. sellers facing their own nonhorizontal demand curves)[7] admits a broad continuum of influence over price. To define the term too narrowly would rule out the investigation of many important situations of seller co-operation with important policy ramifications.

A MODEL FOR PREDICTING COMPANY RESPONSE TO GOVERNMENT CARTELS

On paper at least a mixed, government-business commodity cartel looks like a formidable combination. The Charter of Economic Rights and Duties of States of the United Nations (1974) recognises the right of every state to complete control over the 'possession, use and disposal' of 'its wealth, natural resources and economic activities'.[8] In some commodities, this gives a handful of states control over access to primary mineral products. For their part, the MNCs in industries characterised by oligopoly, may possess enormous market control resources. Indeed, the MNCs may have experience at cartel operation from previous time periods as in the case of oil and aluminium.

Those who argue that the oil companies have colluded with OPEC would accept the logic of just such a proposition. However, there is also evidence that MNC hositility can doom a producers' cartel attempt. In 1974 the banana producing states of Central America attempted to form a government cartel, UPEB. In that case the banana companies 'fought back with noted ferocity'.[9] The corporations engaged in such conflictive acts as cutting production, organising boycotts, and paying bribes to high government officials (United Brands in Honduras). Indeed, the intransigence of their behaviour played a major role in the collapse of that cartel effort. Clearly, it is important that any governments contemplating a producers' cartel be able to predict the response of the MNCs in the affected industry.

Table 1.1 suggests one method of predicting company response on expectations of profit. The basic premise behind this table is that the MNCs are primarily motivated by profit-maximisation and will therefore calculate their response to the proposed cartel on the basis of expected benefits or costs to them. The theory represented by the table

Table 1.1 Result of cartel formation under given economic conditions

Economic condition	*Impact of cartel on profits*	*Responses of companies*
Demand elastic and competitive	Decline	Conflictive
Demand inelastic and fairly competitive	Constant	Passive
Demand inelastic and uncompetitive	Constant or Increase	Passive or Co-operative

can be best explained by applying it directly to the case of bauxite governmental cartel.

The case for calling the International Bauxite Association (IBA) a cartel will be examined more fully later in this study. For my purposes here the important point is that in 1974 most of the IBA members, lead by Jamaica, imposed tax levies on the extraction of bauxite which had the effect of more than doubling the price of bauxite. In such an instance, my expectation would be that the companies would respond according to the effect of the price increase on their earnings for the final product, aluminium.

The effect of the bauxite tax on aluminium depends upon the conceptualisation of the aluminium industry. If, on the one hand, one views the market for aluminium as being elastic and very competitive, one would expect earnings to decline. The aluminium firms would be as middlemen caught between rising costs and a fixed, given market price for aluminium and aluminium products. If, on the other hand, one views the market for aluminium as inelastic and fairly competitive, one would expect earnings to remain farily constant, as the firms would pass on cost increases in the price for aluminium. Finally, if one views the market as inelastic and uncompetitive, one would expect earnings to remain the same or even rise as the firms could pass on the increase and then some.[10]

I can also suggest behaviour patterns of company response for each of the three conditions listed above. In the first situation, I would expect the firms to have an antagonistic attitude toward the cartel and perhaps adopt a *combative* posture toward cartel demands. If the second was the case, I would expect the firms to adopt a fairly *passive* attitude towards the cartel. And in the third instance, I would expect the firms to adopt a passive, *co-operative*, or even protagonistic attitude. However, on the basis of the OPEC price increase, I cannot assume that the firms have perfect knowledge; hence they may go through a learning process of seeing for themselves what happens to their earnings after a one or two fiscal year time lag. This lag may be shortened if the firms have already learned lessons from the OPEC experience which they apply to their own situation.

The purpose of this study, then, is to test this model of MNC behaviour in situations of actual government cartels in primary products. The book will focus primarily on the aluminium industry and secondarily on uranium. Chapter 2 will survey the aluminium industry, describe relevant aspects of the processing of bauxite, and the uses of aluminium, and identify the main actors in the industry. Chapter 3 will demonstrate historically the lack of competition in the aluminium industry. The progressive development of cartel structures and various 'wars' and 'truces' in the industry will be examined. Together Chapters 2 and 3 will establish that the underlying economic conditions for the bauxite/aluminium industry are demand inelasticity and lack of competition: hence the third row of Table 1.1 is most applicable to the bauxite govermental cartel.

Chapter 4 will focus on the pre-bauxite cartel relations between the aluminium majors and Caribbean bauxite produces. It will chronicle the growing frusration of Jamaica and other producers leading to the Guyanan nationalisation and the formation of the bauxite cartel.

Chapter 5 will present the research design and findings. This material uses a Quasi-Experimental design to show that aluminium profits were not hurt by the intervention of the bauxite levy and in fact were outstanding compared to the copper industry benchmark.

Chapter 6 will describe the formation of the producers' governmental cartel, the actions taken by Jamaica against the aluminium majors and the companies' resistance and final acceptance of the new regime. This chapter provides a descriptive complement to the statistical results already presented. The more journalistic evidence shows that the companies changed their pricing strategy on aluminium shortly after the producers introduced the levy. As a result, profits increased in the late 1970s for the firms. The meaning of the phase 'profit sharing' in a 'tax collector' cartel is demonstrated and its impact on various actors assessed.

And finally, Chapter 7 summarises the study, examines cartel situations in two other industries, explains recent events in the aluminium industry and makes projections. In particular it shows how the profit-sharing arrangement is being slowly eroded by (a) the absence of important bauxite producers (Australia and Brazil) and (b) new developments in the industry increasing competition.

2 The Aluminium Industry: A Descriptive Profile

The strategic importance of the oil industry to the world economy has been well established.[1] By comparison to oil, the aluminium industry is largely unheralded. The struggles of the bauxite cartel have received scant media attention though, as this chapter points out, the industry is of crucial significance to the economies of all the AISs and of several Third World states.

The bauxite/aluminium industry is significant for a number of reasons. First, there is the critical importance of the product itself: aluminium ranks second only to iron among metals in world tonnage produced. Aluminium production far outstrips copper, yet the latter has received more attention from writers in political economy. Second, aluminium is a strategic military commodity due to its many nautical and aerospace uses. It is an input in the production of many other manufactured goods (see below) and hence vital to the health of the economies of the capitalists AISs, none of whom are self-sufficient in bauxite. Third, the industry itself is highly concentrated (even more so than oil) both vertically and horizontally, as is most other MNC activity. The industry provides a typical case study in classical MNC activity. While manufacturing rather than mining is the dominant business of most MNCs, a large company like ALCOA, vertically integrated from mining to the fabrication of finished products, engages in both. Finally, individual aluminium MNCs play important roles in the economic and political life of several Third World states. Bauxite, the raw material from which aluminium is produced, is mined in developing areas scattered throughout the Caribbean, Africa and Asia.

In 1974, the International Bauxite Association (IBA) formed a commodity cartel to increase the price of bauxite to the companies through a substantial levy. The industry provides a case study in cartel to company negotiation and hence has been selected for this study to test the theoretical model of cartel formation proposed in the previous chapter.

Before discussing cartel power in the industry further, it is necessary to know something of the processes and uses of the particular commodity. This chapter provides this description along with a brief profile of the major corporate actors.

THE USES OF ALUMINIUM[2]

A 1939 issue of *Life* magazine described aluminium as 'the theme metal of the 20th Century'. The dramatic increase in aluminium production since then has fulfilled that prophecy. Aluminium was then ranked fourth in total metal production: it is now ranked second, behind only iron. Whereas steel production expanded at roughly the rate of GNP increase, aluminium grew three times as fast in the decade following the Second World War. Much of this expanded consumption can be explained in terms of the unique properties of the metal:

Light weight – aluminium has one-third the weight of copper, brass or steel. Because of this characteristic, aluminium is ideal for aviation and other transportation uses. Expansion in aircraft output alone explains much of the dramatic increase in production.

Corrosion resistance – unlike iron and steel, aluminium is not afflicted by rust, thus opening up nautical and architectural uses. These first two properties lead to uses in light boats, aluminium ladders, outdoor equipment and aluminium oxide coatings for buildings. The military has long been a major consumer which explains major production expansions during the wars of this century.

Electrical conductivity – aluminium has twice the conductivity on a pound by pound basis (but not volume) of copper. This property results in aluminium long distance transmission lines. However, the two metals cannot be mixed. As most household wiring has traditionally been copper, aluminium cannot easily replace copper domestically.

Nontoxicity – aluminium gives off no toxic chemicals leading to numerous cooking and packaging applications, aluminium cans, pots and pans, bottle caps and foil.

In addition aluminium finds its way into paints and pharmaceuticals and even fibres in gowns. Aluminium has a low neutron cross section, does not interfere with nuclear chain reactions, and hence has nuclear applications.

The Processing of Aluminium

Aluminium has one of the most complex processing chains of all metals. Vernon (1971) suggests the huge economies to scale in the refining process have been the 'trump card' of the aluminium industry explaining the persistence of vertical and horizontal integration. This process

Figure 2.1 Stages in aluminium processing

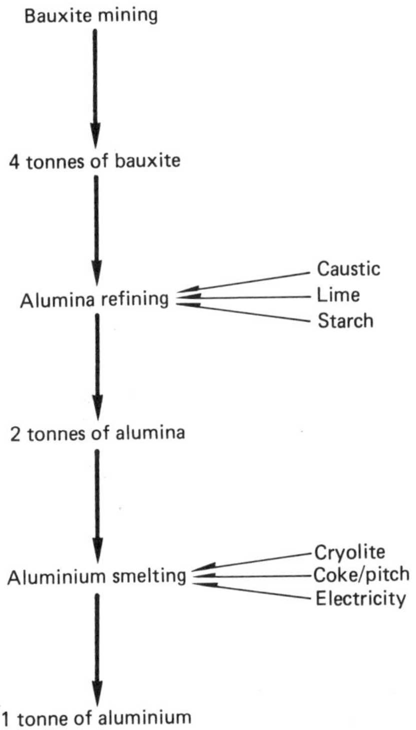

involves five major stages: mining, refining, power, production, smelting and fabricating.

Aluminium is the third most abundant element on the face of the earth, but freeing it from its numerous mineral combinations has been no mean feat. Most aluminium is found in a hydrated oxide form with numerous impurities; that is, the oxide, alumina (A1203) and water are intermeshed with other minerals such as silica and iron oxide. Historically the low-cost source of alumina has been an ore called bauxite. Bauxite ore can be classified as to the amount of water bonded to the aluminium: monohydrates (A1203–H20) such as Boehmite and trihydrates (A1203–3H20) such as Gibbsite. Due to technical specifications of the refining process, Gibbsite is the preferred ore. In addition, impurities and the amount of silica are lower in Gibbsite. Whereas Boehmite is the common bauxite of Europe, Gibbsite is found primarily in the tropics. Surinam, for example, is principally a Gibbsite source. As a result there has been a steady movement of the mining stage away from Europe and the US toward the tropics.

Table 2.1 Leading bauxite producers 1979 (figures in millions of short tons)

	1979 production	*1979 estimated reserves*
Australia	30.4	4959
Guinea	13.5	7163
Jamaica	12.7	2204
USSR	7.2	331
Surinam	5.3	540
Guyana	3.7	771
Yugoslavia	3.3	441
Hungary	3.3	331
Greece	3.2	771
France	2.2	under 99
World total	97.0	24 982

Note: Brazil's production for 1979 was 1.8 but jumped to 5.1 in 1980 which placed it sixth. In 1979 its estimated reserves were 2755.)

Data from Beck *et al.* (1982).

Prior to the Second World War most bauxite mining took place in the developed states though operations were begun in British and Dutch Guiana in the 1920s. The Second World War did much to boost Third World sources and by 1948 the Guianas had 49 per cent of the total. The 1950s saw the rapid development of Jamaica as a source followed by Guinea after 1960.

The next stage is the conversion of the bauxite into alumina. Through a refining operation called the Bayer process, the impurities and water are removed. The fourth stage – reduction – requires copious amounts of electrical power. Hence, as a third stage, a source of cheap electricity, either hydro or thermal, must be secured. Then the reduction of the alumina into aluminium takes place. Oxygen is liberated from a molten bath of alumina with a strong electric current, leaving behind metallic aluminium which is poured into ingots.

Finally, in the fabricating stage, the aluminium is rolled into sheet, extructions, wire and foil. Many aluminium multinationals also engage in the manufacture of aluminium intermediate products and even finished consumer goods. The prevailing technology has moulded the structure of the industry around a few crucial points.

Table 2.2 Leading aluminium producers 1979 (figures in thousands of tons)

	1979 production	*% of world*
US	5023	50%
USSR	1930	19%
Japan	1114	11%
Canada	970	10%
West Germany	818	
Norway	727	
France	435	
China	400	
UK	396	
Italy	297	
Australia	297	
.	.	
.	.	
.	.	
Brazil	262	
India	233	
Venezuela	228	
Ghana	186	
Surinam	71	

Data from US Bureau of Mines

1. *Economies-to-scale* – sound technical reasons favour large-scale processing. Bauxite refining in particular is a large-scale continuous process. The costs in reduction favour at least a 200 000 ton capacity.
2. *Capital intensiveness* – the process requires large capital investments, even more so than in steel making.
3. *Electrical power needs* – electricity is required in such large amounts that it becomes a major component of costs. Hence, the companies must negotiate large, low-priced power contracts from utilities or else acquire their own sources of power. As a result of numerous water power projects along the St Lawrence and Tennessee rivers, ALCOA is itself in certain localities a utility company.
4. *Internationalisation* – the very nature of the technical demands of aluminium-making has forced the internationalisation of the firm. Bauxite must be mined in the tropics, alumina reduced near cheap electrical sources, and aluminium fabricated near consumer markets. Hence a typical European firm mines in Africa, reduces in Norway, and fabricates in France or central Europe close to its major markets. Furthermore, since it is cheaper to transport alumina than bauxite, there

Figure 2.2 Map of aluminium and bauxite producers

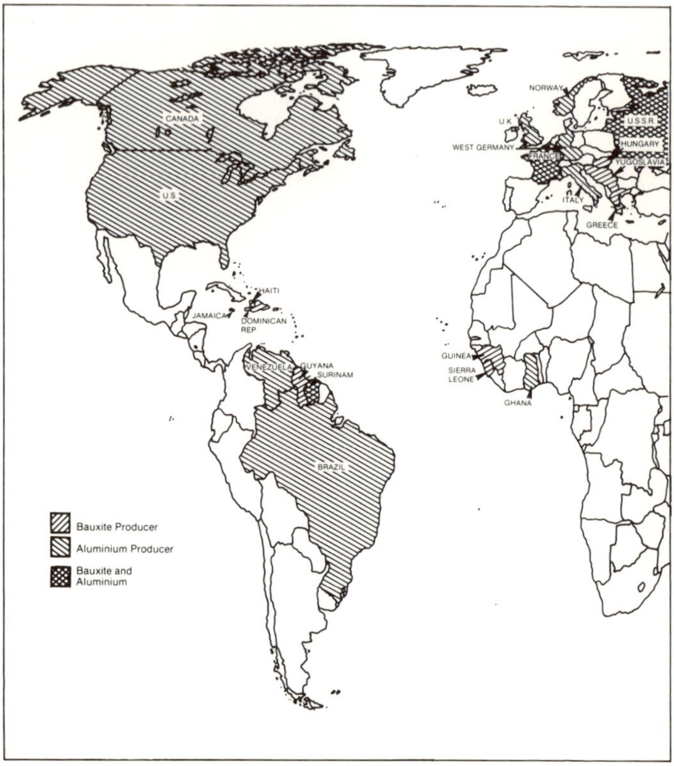

has been a gradual shift of alumina refining to the less developed nation bauxite sources.

In total these characteristics lay the groundwork for describing the industry today. It is an industry dominated by six major integrated international firms with control of the major share of everything from bauxite to fabrication. Before tracing this development historically, let us briefly compare the aluminium processing chain with that of a related metal – copper.

Comparison with Copper Processing

Whereas the theme of aluminium's story has been the long search for a way to produce it economically, copper was being mined extensively in ancient times. The Egyptians used copper pipes in their plumbing; the Romans used it in their architecture. Copper was originally mined from veins so rich in the metal that little processing other than smelting was needed. But with time the purer veins have been exhausted. Hence the search has been for technologies to process increasingly impure forms of the mineral.

Today copper production involves these major stages – mining, smelting and refining – though the exact processing varies with the specific ore. The 'refining' and 'smelting' stages are reversed from those of aluminium; refining here is the final removal of miniscule impurities from the smelted copper. With western US ores this refining is done with an electrolytic process.

Narvin (1978) states that whereas aluminium reduction is the dominant stage in aluminium production, mining is the dominant stage in copper production, due in part to the complexity of ore types. In general, the copper majors have not integrated as far forward (marketward) as the aluminium companies; few copper majors own fabricators, for example. The energy requirements are nowhere near as intense as in aluminium, but copper mining is very capital intensive. The physical assets per worker can be as high as $100 000.

The copper industry has already attracted the attention of many scholars. For greater descriptive material and analysis, the reader should see Moran (1974) and Brundenius (1972).

THE HISTORY OF ALUMINIUM[3]

Aluminium's first great benefactor was no less august a person than Napoleon III. Attracted by tales of the elusive metal's light-weight strength, he envisioned casting his French legions in aluminium armour. But unfortunately for the Emperor, the abundant aluminium oxide refused to release its metal component easily. By the mid-nineteenth century small laboratory amounts of the metal existed but, at $115 a pound, aluminium was consigned to use as jewellery. Under the patronage of his Emperor, St Claire Deville developed a chemical reduction process by 1856. But while the Deville process made commercial production possible, at $115 a pound, Napoleon III was forced to abandon his martial vision and content himself with being the only European monarch with his own set of shiny aluminium silverware. None the less the Emperor's active interest put France into the aluminium race; the firm set up to exploit the Deville process was to develop into one of the world's major aluminium MNCs – Pechiney.

It was to be a few decades before simultaneous discoveries in the US and France allowed cheap commercial reduction of alumina to aluminium. Each of the processes involved the same electrometallurgical technique used even today, but resulted in a miasma of patent agreements. The manipulation of these patents set the basic structure of the infant industry. The French Herout patent (1886) resulted in the creation of two firms: a French firm to exploit the process locally and a Swiss firm to process aluminium to the rest of the world. The *Société Electrometallurgique Français* (Froges enterprise) entered into competition with the older Pechiney firm (still using the Deville process.) German backing played an important role in the Swiss *Metallurgische Gesellschaft* (1887) renamed first in 1888 Aluminium Industries AG (AIAG) and then in 1963 Swiss Aluminium (Alusuisse).

In the US the Pittsburgh Reduction Company (1888) emerged from early patent battles as the sole producer of American aluminium. Credit for the American version of the electrolytic process went to Charles Martin Hall of Oberlin, Ohio. Hall with the backing of Captain Alfred E. Hunt, and shortly thereafter the Mellon family, began reduction in Pittsburgh but was soon besieged by patent cases. The first patent challenge came from the Heroult patent but the US Patient Office found in Hall's favour in 1889. The second patent challenge was from the older Cowles Brothers Electric Smelting and Aluminium Company which held a shotgun Bradley patent on any 'heated electric arc' aluminium process. In an out-of-court settlement (1893), Pittsburgh Reduction paid Cowles $1.4 million to stop producing aluminium. This is was that Pittsburgh Reduction (or, ALCOA, as we know it today) gained a domestic monopoly which it was to retain to the end of the Second World War.

Some of the historical activities of these firms – ALCOA, Pechiney, and Alusuisse – will be recounted elsewhere. At this time it should be noted that, on the basis of the technical aspects of the industry cited above, the industry has come to be dominated by six large, vertically integrated companies: ALCOA, Reynolds, Kaiser, ALCAN, Pechiney, and Alusuisse. Profiles of each of these firms follow.

Alcoa

Though its market share has declined slowly since the Second World War, ALCOA is still the world's largest aluminium producer with

1 575 000 tons per year capacity in 1974 in its American plants alone (Aluminium Association, 1974, p. 32). In 1975 (which was not a banner year for metal sales) ALCOA ranked only 85th by sales of American corporations (it was 64th in 1974) but with over $3 billion in assets, it ranked 36th on the Fortune list, putting it in the same league as Proctor & Gamble and International Harvester. ALCOA had to rebuild its overseas operations after the 1950 loss of Alted, which became ALCAN (see ALCAN below). At present it has major bauxite mining operations in the US, Norway, Australia, Mexico and Surinam. The Surinam metal gives it duty-free access to EEC countries. It has a strong position in the US fabrication industry. Though originally built on hydroelectric power, ALCOA's new plants on the US Gulf Coast use thermal energy. The company's greatest asset is its strong domestic position in sales and production though, through subsidiaries in Norway, Surinam and Japan, it has grown into all the major developed markets. ALCOA is also a leader in new industry sales and technology.

Reynolds

Reynolds is the world's second largest aluminium firm and a major marketer outside the US. The most important step in the internationalisation of the company was the acquisition of a 51 per cent share of the older British Aluminium Company (46 per cent directly, 5 per cent through another affiliate). This move gave it a strong position in the UK, Canada, and Norway. The firm has bauxite sources in the US, Jamaica, Guyana, Haiti, and is a minor participant in both the Fria Consortium (Guinea) and the Ghana Consortium. It owns fabricators in the US, Canada, UK, Belgium, Japan, Germany and the Philippines. Reynolds is acknowledged as the leader in packaging uses of aluminium.

Kaiser

Kaiser is the most diversified of the majors with 79 per cent of its sales in aluminium. It also has interests in refractories, chemicals, fertilisers, and nickel. Kaiser was unable to acquire bauxite sources in the US and hence has a very bold, international location policy. It has often been called the most sensitive to Third World host nation situations and is the only American firm with a major African interest Ghana. In the late 1960s Kaiser began heavy investments in Australia.

Alcan

This Canadian firm has had a long history of association with the American giant of ALCOA through ALCAN's many incarnations. Its first 'life' was begun in 1900 as ALCOA's directly-owned Canadian subsidiary – Northern Limited – set up to control the former's Canadian smelters and access to Canada's cheap hydroelectric power. Then, as will be seen, it was reborn in 1928 as Alted, a 'holding company' for ALCOA's interests. In 1950 it began its third life with Judge Knox's decision that removed most of the Mellon interests, though it did not adopt its present name until 1966. ALCAN is called a Canadian firm but in reality, as Brubaker (1967) points out, 70 per cent of its stock is owned by Americans.

Given that the firm's structure was historically set by the needs of ALCOA's international strategy, ALCAN finds itself in the tenuous position of being a huge far-flung international business with a weak domestic position. Its share of the Canadian market is simply not enough to absorb much of its Canadian capacity of 950 000 tons per year. Hence, in contrast to ALCOA, ALCAN does not have a strong domestic base to fall back on in periods of international aluminium slump. At first it supplied much aluminium to independent US fabricators, but in the late 1950s the big American firms began acquiring these independents and weakened ALCAN's position. ALCAN retaliated by acquiring American fabricators of its own and further diversifying its international markets.

ALCAN is a fully-integrated firm with bauxite sources in Jamaica, France, Malaysia, Sarawak and, until recently, Guyana. It has alumina plants in Quebec, Jamaica, Norway, Australia, Brazil, India and Japan. It has smelting and fabricating operations in 30 countries including most of the European countries, Argentina, Brazil, India, Japan and Australia. Thus the firm is the most international aluminium MNC, which at times contradicts its pledge to Canadian production. Its strengths include its position as low-cost producer (due to intensive Canadian Norwegian hydroelectric properties), its high technology base, and its international expertise.

Pechiney

Pechiney is the oldest aluminium firm tracing its ancestry back to the 1859 St Clair Deville process. The firm has had a strong domestic base in France and now in the EEC. It obtains bauxite from affiliates in

Southern France, Guinea, Australia, and Greece, and owns alumina refineries in the same nations. Its smelters are located in France, Cameroon, Spain, and the US (through its 1975 acquisition of Howmet).

It should be remembered that France with the US was a birthplace of the aluminium industry. By the time of the Aluminium Alliance (see Chapter 3) there were three major aluminium firms in France: Pechiney, Froges, and Société d'Ugines. At some point since, Pechiney (based on the lists of present affiliates) absorbed Froges. Then, in December 1971, in one of the major European mergers of the century, Pechiney absorbed its remaining domestic rival – Société Ugine-Kuhlmann – to become Pechiney Ugine Kuhlmann, the world's fourth largest producer and one of the largest industrial combines in Europe. The new firm is highly diversified with business in copper, steel, rare metals, dyes, chemicals, pharmaceuticals, and nuclear technology.

Alusuisse (Swiss Aluminium A.G.)

Alusuisse is now the sixth largest firm and the only one without a domestic market of its own. Alusuisse owns bauxite facilites in France, Italy, Greece, Sierra Leone and Guinea through a minor interest in the Fria Consortium. Most of its refining, smelting, and fabricating operations are located in Italy, France, Germany and Switzerland. Swiss Aluminium entered the US market with an American affiliate, Conalco in 1948. In 1963 it constructed a reduction plant at New Johnsonville, Tennessee.

The Other American Producers

Since the 1950s other firms have entered the aluminium business in the US, but the big three still own two-thirds of the nation's primary capacity. Most of the minors are not fully integrated and must rely on the majors for purchases of bauxite for alumina. Nearly all of the minors have had complex changes in ownership and are hence difficult to trace.

Harvey-Martin Marietta

The story of the intrepid Harvey Aluminum Corporation will be told in the next chapter. In 1968 Harvey fell under the acquisitive sight of Martin Marietta, a chemical aerospace conglomerate, which bought up 41 per cent of its stock. Within a few years, Martin Marietta purchased more blocks of stock, changing Harvey's name to Martin Marietta Aluminum in 1972, and absorbing the firm completely in 1974.

Anaconda Aluminum

Due to the troubles of its parent, this firm has never completed its plans for vertical integration. In January 1977 it disappeared with its parent into the Atlantic Richfield Company.

Phelps Dodge Aluminum

Other copper companies have attempted to diversify into aluminium like Anaconda, on the whole with unsatisfactory results. The Phelps Dodge Copper company made such an attempt but in 1971 sold its subsidiary to CONALCO (Swiss Aluminum) for a 40 per cent interest in that firm.

Ormet

This was a joint – Olin, Revere Copper and Brass – venture. This partnership was sundered by Olin's January 1974 decision to get out of aluminium. Olin sold its share to CONALCO (Swiss Aluminum) and Revere retained its 34 per cent share.

Intalco

This consortium was formed in 1966 by Pechiney, Howmet, and AMAX. It apparently was rather short-lived. In 1974 AMAX sold 50 per cent of its meagre aluminium interests to Mitsui of Japan. In 1975 Pechiney acquired Howmet and absorbed the remaining aluminium interests directly into Pechiney Ugine Kuhlmann.

Other Minor Producers

VAW is the large German state-owned firm (218 000 ton capacity in 1969), but since the war has shied away from international business. The Italian Montecatini-Edison splits its domestic market with Alusuisse.

Japan has produced two important independent firms: Showa Denko K.K. (SDK) and Sumitomo Chemical Company. Both firms draw their bauxite from Malaysia, Indonesia and Australia.

SUMMARY

While the above profile lists many firms involved in the processing of aluminium, it should not be forgotten that the industry is in fact dominated by six 'majors'. The domination of the three American firms over the largest capitalist economy is a textbook example of oligopoly (see Samuelson, p. 482.) The international industry today can be best described as an oligopoly with a fringe: the six majors (ALCOA, ALCAN, Reynolds, Kaiser, Pechiney and Alusuisse) comprise the oligopoly with all other firms described above relegated to the fringe. This characterisation is supported by the fact that as recent as 1970 the majors controlled 80 per cent of US aluminium capacity and a similar share of all market-economy nations' capacity.

In searching for explanations for this oligopolistic structure, a number of factors are evident. The importance of control of technology can be seen by the fact that three of today's majors were established out of the original copyright battles over the Herout and Hall reduction processes (ALCOA, Alusuisse and Pechiney). It could be argued that the remaining three exist today only through the intervention of the Anti-Trust section of US Justice Department and the Truman Administration's desire to break the fifty-year US monopoly of ALCOA.

The fact that the nineteenth century originators have prospered so long suggests huge economies-to-scale and high barriers to entry for would-be competitors. As has been shown above, a complex processing chain, capital-intensiveness (high capital start-up costs), energy requirements, vertical integration and internationalisation to control bauxite sources and markets have all played a part. Life in the fringe is quite risky as can be seen from the company profiles. Assets turnover rapidly and even large companies have left the field not long after entering. In the US market, would-be competitors have taken the form of larger consortiums though even these have frequently been dominated by other majors (Alusuisse with Phelps Dodge and Ormet, Pechiney with INTALCO).

It should not be surprising, then, to find that the majors have made use of their structural advantages to create business cartels at various times in the past. Chapter 3 investigates these past arrangements to dominate world markets.

3 History of the Aluminium Industry's Cartels

The contemporary highly interwoven pattern of connections among the aluminium multinationals has a long history stretching back to 1886 and the simultaneous discoveries in the US and France of a cheap commercial process for the reduction of alumina into aluminium. Both of these processes involved the same electrometallurgical technique used even today but at the time it resulted in a miasma of patent agreements. As noted in the previous chapter, the manipulation of these patents set the basic structure of the infant industry. The French Herout patent (1886) resulted in the creation of two firms: a French firm to exploit the process locally and a Swiss firm to process aluminium in the rest of the world. *The Société Electrometallurgique Français* (Froges enterprise) entered into competition with the older Pechiney firm. German backing played an important role in the Swiss *Metallurgische Gesellschaft* (1887) renamed first in 1888 Aluminium Industries AG (AIAG) and then in 1963 Swiss Aluminum (Alusuisse). In the US, the Pittsburgh Reduction Company emerged in 1888 from early battles as the sole producer of American aluminium under the Hall patent. This company later became the Aluminium Company of America (ALCOA) in 1907.

In this chapter the development of 'international diplomacy' among these firms will be traced. An investigation of the MNC in the contemporary bauxite cartel would not be complete without examining the use of cartel power in the industry historically. Cartels are not new to the bauxite/aluminium industry. Thus this section will focus on the interaction between ALCOA and its international rivals from the nineteenth century to the present. As will be seen, this period reveals a striking array of co-operative and conflictive behavioural patterns; the primary issues were prices and markets.

ALCOA'S 'FOREIGN' POLICY[1]

By the time of the Pittsburgh Reduction Company's metamorphosis into the Aluminium Company of America (ALCOA) in 1907, it was already a fully-integrated firm both vertically and horizontally. With sole control of the Hall patent and the capital resources of its Mellon family owners, it was to enjoy a monopoly on aluminium production in the US down through the Second World War. Vertically, the company owned everything from the bauxite lands in the southern US to the Aluminum Cooking Utensil Company which distributed a finished product.

Having successfully pre-empted domestic competition, ALCOA's main business threats came from overseas and went back to the nineteenth century. The Swiss AIAG and French Froges grew at the same pace as ALCOA and in 1895 the British Aluminium Company was likewise licensed under the Heroult patents. The ever-growing large American market attracted them all in spite of an eight cents per pound tariff. ALCOA moved immediately to deal with the foreign threat, first by leasing its own process to Froges' rival Pechiney (1895) and second by reaching agreements (1896) with AIAG not to export to each other's markets.

This agreement was just the first in a series of cartel agreements which were to dominate world aluminium for the first half of the twentieth century. The second cartel, signed on 2 November 1901, was more comprehensive than the 1896 pact; it included AIAG, ALCOA, British Aluminium, Froges and Pechiney. It closed the respective home markets of Switzerland, the US, Canada, the UK and France, and assigned quotas on the 'open markets' (ALCOA's share of which was 21 per cent). The cartel's governing body fixed prices for these open markets. The closed market price was to be kept 1¢ higher to prevent independents from buying locally and selling on the open market. The result of cartel policy was a price advance to 36¢ per lb. in Europe. Through a prior agreement ALCOA kept the US price at 33¢ until 1906 when it conformed to the cartel price. In doing so it doubled its productive capacity from 1901 to 1906. Its rates of return also rose with the price increase (see Table 3.1).

The third cartel, formed in 1906, was to have been a renewal of the previous agreement with tough enforcement and regular reports from

Table 3.1 ALCOA's net profits on stockholder equity

1899	11%
1900	15%
1905	26%
1906	35%

Source: Stocking and Watkins (1964).

members on sales, but a combination of deleterious events blighted this pact. First, the Heroult patents expired between 1906 and 1908, opening the European field to competitors attracted by the high profits. Second, there was a general business downturn in 1907. Finally the 'high price' policy of the cartel disenchanted some of the original members who believed that lower prices were needed to create markets for the still relatively new product. The members dissolved this short-lived cartel in 1908.

A similar short-lived fourth cartel agreement between ALCOA and AIAG failed to curtail a burgeoning aluminium invasion of the US. Even with a 7¢ tariff, ALCOA's overseas competitors exported 7.6 million pounds to the US in 1910. The US price dropped 22¢ per pound, but a dramatic rise in domestic consumption more than compensated the US firm.

The fifth cartel, proposed by ALCOA President A.V. Davis on a trip to Europe in 1911, was founded upon the realisation that ALCOA could no longer preserve the US market as its private domain. But the 1912 anti-trust case of the US Justice Department complicated this agreement: ALCOA had promised not to involve itself directly in any further Euorpean cartels. At this point ALCOA established the expedient of participating in cartels through its Canadian subsidiary, North Aluminium. ALCOA signed the Consent Decree with the Justice Department on 7 June 1912 and five days later signed the cartel agreement. The Democratic administration of 1912 lowered the aluminium tariff to 2¢ a pound and imports rose to 20 million lbs. in 1913. But ALCOA proposed to create a new American firm, Southern Aluminium, under joint ALCOA and European control. It hoped to use this ploy as a safety valve for European desires on the US market. The outbreak of war intervened to end both the project and the cartel.

The First World War dramatically raised demand for aluminium products, such as helmets, canteens, aluminium shells, explosives (a dust with ammonium nitrate), mess containers, and, most importantly, airplanes and engines of all types. Consequently all firms doubled their productive capacities and new plants sprang up in Norway and in Germany (the German state monopoly, VAW). After the war ended, the aluminium price war began in earnest as European firms dumped their stockpiles on the American market.

THE POST-WAR ALCOA OFFENSIVE

ALCOA responded to this post-First World War 'invasion' with a tough offensive. On the domestic front it lobbied for the 1922 Fordney-

McCumber tariff which raised the duty to 5¢. ALCOA then attacked its European rivals directly with an aggressive acquisition programme aimed at forcing a new cartel. The new Norwegian firms were the main source of cheap aluminium imports. In 1922 ALCOA purchased 50 per cent and control of Norsk Aluminium, and in 1923 purchased a third of *Det Norske Nitrid.* In 1925 it acquired half of *Societa dell' Alumino Italiano* and a large portion of AIAG's *Alumino Espanol SA*. After vainly pursuing a block of VAW stock, it settled for an agreement with the German firm not to export to the US.

Another element of its strategy involved buying up sources of European bauxite. In France, the Société des Bauxite Françaises held its bauxite properties. Yugoslavia was the main new source of bauxite for Europe and in rapid succession ALCOA acquired *Jadranski Bauxit* (1922), *Primorski Bauxit* (1925), and *SA Mineraria Triestina* (1926). Finally ALCOA sought leverage over its rivals by purchasing their hydro electric sources: in 1924 55 per cent of *Det Norske Aktieselskab Electrokemisk* and *AS Kinservik* (Norway).

The European companies sued for an armistice in the informal sixth cartel (1923) which contained a gentleman's agreement on price and individual agreements to limit exports to the US. Table 3.2 dramatises the impact on aluminium prices. Meanwhile, demand for aluminium expanded rapidly through the 1920s in the automotive and aviation industries. Profits began to revive and ALCOA again increased its capacity.

Table 3.2 Aluminium price per pound before and after sixth cartel

	1922	*1924*
Europe	18¢	24¢
U.S.	19¢	27¢

Source: Stocking and Watkins (1964).

The Seventh Cartel (1926) formally cemented the peace along the old pre-First World War lines. It re-established quotas on all sales, domestic and export, penalties for oversales and compensations for undersales; it made standard delivery prices uniform for all world markets. Members submitted quarterly accounts on sales volumes, prices, output and inventories to the governing body. Due to US anti-trust laws, ALCOA again participated through its Canadian subsidiary. None the less, ALCOA's indirect participation in the cartel can be seen in its strict policy of refusing sales to non-cartel member firms. The cartel was to have lasted two years, but at the end of 1928 it was extended three more years. The ancillary Zurich Agreement (1930) divided the new Japanese market among the members (ALCOA received 52 per cent through its Canadian branch) and later agreements similarly divided India and Russia.

As the cartel agreements became more intricate, ALCOA needed a more direct way to participate without violating the 1912 Consent Decree. On 31 May 1928 it obtained a Canadian charter for its reorganised Canadian subsidiary (renamed ALTED) and transferred to it nearly all its foreign holdings; only the critical bauxite holdings in Surinam were kept with the parent. In effect, ALTED became a holding company for ALCOA, but to maintain the image of independence, ALTED's stock was distributed to ALCOA's shareholders on a pro rata basis. Over the years a few names changed, but not enough to threaten ALCOA's control and certainly not enough to justify the claim that ALTED had 'no connection at all with the American company'.[2] In reality the same interests (Mellons, Davises, Hunts) dominated the stock of ALTED. Indeed the President of the new firm, E.K. Davis, was a brother of A.V. Davis, President of ALCOA. ALCOA continued to operate in tandem with ALTED and through it with the cartel. Ultimately, in 1950, the courts were to trap ALCOA in its own fiction and so force a separation of the two.

THE ALUMINIUM ALLIANCE

On the eve of the depression the ALCOA-ALTED combine controlled around 50 per cent of world production. But the onset of the depression threatened all aluminium firms. In 1930 A.V. Davis again travelled to Europe and again a new cartel agreement emerged. The business downturn called for drastic measures and the new pact represented one of the greatest innovations in cartel history. In effect, the new deal 'incorporated' the cartel; every major aluminium firm signed the Foundation Agreement in Montreal in 1931 to form a super-firm, Aluminium Alliance. Officially incorporated in Switzerland, the Aluminium Alliance issued 1400 shares of stock to members only on the basis of one share for every hundred metric tons of capacity. ALTED received 400 shares or 28.6 per cent, the French firm 21 per cent, the German 20 per cent, the Swiss 15.5 per cent, and the British 15 per cent (see Figure 3.1). Thus voting power and directorship allotments

Figure 3.1 Alliance Aluminium Compagnie

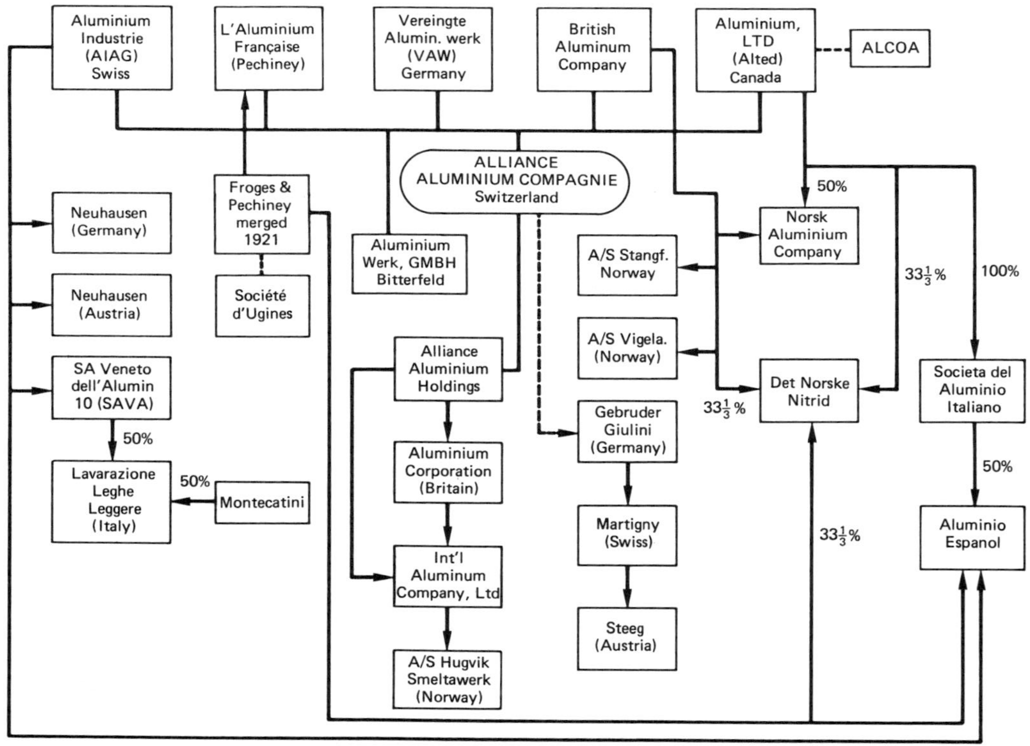

Sources: Information from Stockings and Watkins, and Company annual reports.

reflected the already established market quotas. The Alliance directors set policy, fixed minimum prices, and bought and sold aluminium to bolster the administered price. Once again ALCOA acted as the 'silent partner' operating in accord with Alliance policies. When the Alliance refused to give technical assistance to the Japanese Government which sought to create an independent Japanese aluminium industry, ALCOA likewise shunned the overtures.

The greatest measure of the Alliance's success was that aluminium prices were maintained throughout the depression at stable levels until the various national rearmament programmes made the need for price coordination superfluous. Alliance policy also collided with the German Nazi government which wanted to force VAW out of the cartel. Alliance agreed to remove restrictions on VAW's production so long as that firm agreed not to export its aluminium. With the advent of war, collaboration became unnecessary if not impolitic. Alliance and the war helped ALCOA and ALTED prosper greatly as they were to emerge from hostilities with 64 per cent of world production.

NEW RIVALS FOR ALCOA

Throughout the 'Golden Era' of aluminium cartels (roughly 1896–1945) ALCOA had been able to play a leading role through its monopoly domination of the largest capitalist market for the commodity: the United States. In turn, this monopoly depended on ALCOA's ability to preempt the creation of domestic rivals and to circumvent the anti-trust enforcement efforts of the US Justice Department. Higher prices and profits acted as lure to would-be competitors but, as we have seen in Chapter 2, huge barriers to entry existed to deter those same competitors: patent control, high capital costs, large electricity requirements, and ALCOA's immense economies-of-scale were all important factors.

None the less, the company worried about one of its own customers or suppliers integrating backward or forward into aluminium smelting. One example of this concern is shown in the Justice Department case of 1912 which alleged that ALCOA made restrictive convenants with its suppliers: ALCOA would purchase from them only so long as they stayed out of aluminium. Other contracts and covenants kept alum-makers out of the aluminium industry. In the Consent Decree of 1912 the company agreed to refrain from such activities and to stay out of foreign cartels.

The Consent Decree of 1912 as we have seen did not prevent ALCOA from indirectly participating in the cartels of the 1920s and 1930s through its Canadian subsidiary (later ALCAN). It seemed to entail a warning that the Justice Department was watching ALCOA for anti-trust violations. Yet the next complaint was not filed until 1937, a quarter-century later. To understand why we must consider ALCOA's 'political connections'.

From its early days as the Pittsburgh Reduction Company, the Mellon family had been a supplier of capital and major shareholder in the company. In 1920 a prominent member of that family, Andrew Mellon, accepted the position of Secretary of the Treasury under President Harding, a post he was to hold through the following Republican administrations of Coolidge and Hoover. Mellon's policies had a fiscal conservative, pro-business bent and he actively lobbied for and got the reduction of taxes on the rich and the giving of tax rebates to corporations for 'over taxes' collected during the First World War. William Hoffman (1974) claims to see the hand of Mellon behind the Ford-McCumber tariff on aluminium imports.

Mellon's policies were blamed for the onset and continuation of the great depression in the Democratically-controlled Congress. These accusations led Representative Wright Patman to initiate impeachment proceedings against Mellon in January 1932 on the grounds of conflict of interest. Among the specific counts listed by Patman were:

1. in his role of chief on the Coast Guard, he oversaw the collection of import duties on goods his businesses imported;
2. he gave his own businesses large tax rebates;
3. as *ex officio* chairman of the Federal Reserve Board, it was illegal for him to own bank stock;
4. his distillers sold liquor during Prohibition; and
5. he advised the construction of public buildings using aluminium.

Before the hearings had reached the impeachment stage, however, Hoover rescued that Mellon by appointing him Ambassador to England.

Several of the charges listed above allege that Mellon used his position to benefit ALCOA. It is interesting to note that during the three Republican administrations, no legal action was taken against the ALCOA monopoly. Attorney General Harlan F. Stone was preparing an anti-trust case against ALCOA when Coolidge suddenly appointed him to the Supreme Court. 'Many observers thought they detected the long arm of the Secretary of the Treasury in the convenient removal of Attorney General Stone, especially since his successor never followed up on the charge.'[3]

In the late 1930s, under the New Deal, the US saw a second wave of anti-trust activity. ALCOA was an obvious target and in 1937 the Department of Justice filed a broad complaint of monopolisation. In 1940 District Judge Caffery acquitted ALCOA, but in 1945 Circuit Court Judge Hand found ALCOA guilty of a price squeeze on independent sheet fabricators. ALCOA was charged with raising the price of ingots to the independents while at the same time lowering its own sheet metal price. However, Judge Hand postponed any remedy of the situation until the end of the war.

If the New Deal made a nemesis for ALCOA out of the Justice Department, then the Second World War made an ally for it out of the Defense Department. Indeed, the war had brought ALCOA into direct co-operation with the Defense Department. It became clear to the latter upon America's entrance into the war that ALCOA could not expand its capacity enough to meet war aviation needs without government assistance. In 1942 the Government's Defense Plant Corporation, a subsidiary of the Reconstruction Finance Corporation, began subsidising the construction of aluminium plants. By 1944 government production accounted for nearly half of the country's aluminium capacity, and ALCOA operated four-fifths of these new plants through management contracts and leases. At the end of the war the Defense Plant Corporation turned over its properties to the Surplus Property Board.

The fate of these plants now became a hotly contested issue. ALCOA stood in a good position to claim the lion's share; it had leases on the plants till 1947–48 and had seen to it that the plants were constructed with ALCOA patented processes. Also in ALCOA's favour was the fact that Congress wanted the plants sold as quickly as possible with little interruption in operation. Finally, there was a scarcity of applicants for the plants indicating the trepidation of any would-be buyer at the prospect of competing with the ALCOA combine. Reynolds Metals, which had began producing aluminium during the war, wanted subsidies and guarantees of Government purchases before it would acquire any of the plants.

In July of 1945 President Truman appointed W. Stuart Symington as the Surplus Property Administrator. Symington immediately appointed two economic advisers, Gordon Reed and Sam Moment. These two economists proceeded to draw up conflicting plans for disposing of the properties. Reed recognised ALCOA's strong interests and proposed that the firm be allowed to purchase plants up to the point that it would hold 60 per cent of US capacity.[4] The rest would be sold to competitors. Sam Moment's plan emphasised the creation of effective competition in the aluminium industry. He pointed out that only large vertically integrated firms would be capable of competing with ALCOA. Therefore, he suggested that all the properties be used to set up two integrated firms with the government's two alumina refiners serving as the backbone of each.

Symington vacillated between the two plans before accepting the Moment plan, though he did allow ALCOA to acquire the St Lawrence Reduction works and an extrusion plant at Cressona, Pennsylvania. He solved the leasing problem by cancelling the contracts under the clause that the plants had to be operating above 40 per cent of capacity. ALCOA resolved the patents problem by granting royalty-free licenses. The subsidy plan failed to pass Congress, but a government stockpile programme, beginning in 1945, increased civilian uses, and the Korean war stimulated demand and took up any slack production. Under these conditions Reynolds Metals agreed to purchase the Hurricane Creek alumina plant and the Jones Mills, Arkansas and Troutdale, Oregon aluminium plants. Kaiser Aluminum and Chemical Corporation acquired the Baton Rouge alumina plant and the Spokane and Tacoma, Washington aluminium plants. The plan resulted in three US primary producers with capacity divided: ALCOA 51 per cent, Reynolds 29 per cent, Kaiser 20 per cent.[5]

In 1947 ALCOA petitioned the courts that it no longer was a monopolist. The Department of Justice counter-petitioned that true competition would not exist unless ALCOA was split-up, creating four competitors of equivalent size. In 1950 Judge Knox ruled against the government but turned his attention on ALTED, ALCOA's Canadian partner. Here he found that nine individuals in the Mellon and Davis familes owned 46.3 per cent of ALCOA and 44.7 per cent of ALTED and ruled that these individuals must sell their stock in one or other of the firms.[6]

If, according to ALCOA's past statements, the two were acting independently, the change of stock should not represent any change in operating practices. Caught in its own rhetoric, ALCOA agreed. Knox also put ALCOA on a five-year probation period. In 1957 the courts ended the case. The North American aluminium monopoly had been replaced by four vertically-integrated international firms as a result of direct government intervention.

THE CARTEL AFTER THE SECOND WORLD WAR

The last known official business meeting of the Aluminium Alliance was held in May 1944. In December of 1945, at the request of its Canadian member, the cartel was dissolved. During the war the US and British governments had blacklisted the cartel and the US Alien Property Custodian seized its American assets. With the new wave of anti-trust feeling among North American and West European governments, cartels of the Golden Era could, at the very least, not operate openly. Furthermore, American anti-trust activity had created three newly-independent integrated companies. ALCOA, stripped of ALCAN, had lost most of its international holdings and business. Clearly, ALCOA could no longer exercise of world leadership role as it had from behind the scenes in the past. With its domestic monopoly broken ALCOA was even unable to prevent the non-American companies from invading its home market. Alusuisse formed a US affiliate, CONALCO in 1948 and by 1956 it was building an aluminium smelter in Jackson, Tennessee. Likewise, Pechiney gained an American foothold through INTALCO in 1966. A new period of retrenchment and rivalry seemed to be indicated in the post-war period.

But there were also signs of continuing collusion in spite of the absence of a formal cartel structure such as the Aluminium Alliance. One such indication is the relative stability of the oligopoly of the six majors. Though three of today's majors emerged from the Second World War and after, the overall number of important firms has remained relatively constant due to post-war consolidations. The most important consolidation was the 1959 acquisition of the ailing British Aluminium Company (BAC) by Reynolds. BAC had just completed its own North American invasion with a new smelter at Baie Comeau in the early 1950s. Strapped with a huge debt load and other financial problems, BAC had become a take-over target. In May 1957 Reynolds merged with the large British aluminium fabricator – Tube Investment Ltd. Reynolds' larger strategy became clear when, in December 1958, Reynolds and Tube indicated they were purchasing BAC stock. Despite a brief 'aluminium war' complete with an eleventh hour appeal for ALCOA to rescue it, BAC was 80 per cent Reynolds owned by the next month. Hence, Reynolds gained the Baie Comeau smelter, access to both Australian bauxite and European markets and a world-wide distribution system in one fell swoop.

Other consolidations have involved disappearances of important 'fringe' members. On 1971, Pechiney absorbed its last domestic rival

Ugine-Kuhlmann and in the US, CONALCO (Alusuisse) purchased Phelps Dodge's aluminium company. In 1975, Pechiney acquired its former American partner in INTALCO – Howmett.

The majors have also collectively overcome challenges from ambitious fringe members. The sad tale of Harvey Aluminium is instructive here. Harvey entered the aluminium industry at the close of the Second World War by purchasing a government Defence Plant Corporation fabricator. Next it secured government support through a five-year aluminium stockpile contract. But Harvey ran into the same obstacles facing other would-be aluminium majors. It was lucky enough to sign a long-term power contract with a new dam in Montana but discovered it did not have the capital to complete the smelter it was building near the site. In 1953 it had to sell its interest in the Montana smelter and settle for a smaller smelter plant in Oregon. But in Oregon it had problems negotiating a long-term power contract with the local utility. After litigation and an out-of-court settlement, Harvey got its power but only on condition it build its own power lines. Then, there was the problem of finding alumina to smelt in its Oregon plant. The majors Kaiser, Reynolds, ALCAN and ALCOA each refused to sell alumina to the newcomer. But the intrepid Harvey began importing Japanese alumina.

Harvey's only hope of successfully challenging the majors was obviously to became an integrated international company itself. As will be shown in the next chapter, in 1963 it signed a path-breaking government joint-venture agreement for a bauxite mine in Guinea. Guinea and Harvey went to the World Bank for an infrastructure loan only to be told that Harvey was too small for the project. Under the Bank's pressure, ALCOA, ALCAN and Pechiney were brought in to take the lion's share of the Boke (Sangaredi Plateau) bauxite production. Shortly thereafter, Harvey was brought out by Martin Marietta. As I concluded at the end of Chapter 2, the fringe has yet to successfully challenge the oligopoly.

One might ask if the absence of an 'aluminium alliance' with its formal meetings doesn't rule out the opportunity for price and market collusion. Yet a 1980 study of rates of return suggests that in the US, aluminium firms may be earning monopoly profits.[7] Furthermore, as with the oil industry, the majors are linked internationally through consortiums. Figure 3.2 illustrates these interlinkages. Aside from the consortium linkages, the majors are members of numerous domestic and international industry associations. In the US, there is the Aluminium Association founded in 1933 which today has 91 members (most are small fabricators of aluminium products produced by the majors). Britain has the Aluminium Federation founded in 1962, which included committees on 'international relations', 'statistics', and 'product liability'. Japan and Western Europe have their own associations. More important is the International Primary Aluminium Institute (IPAI)

Figure 3.2 Inter-linkages of consortiums in the bauxite/alumina/aluminium industry

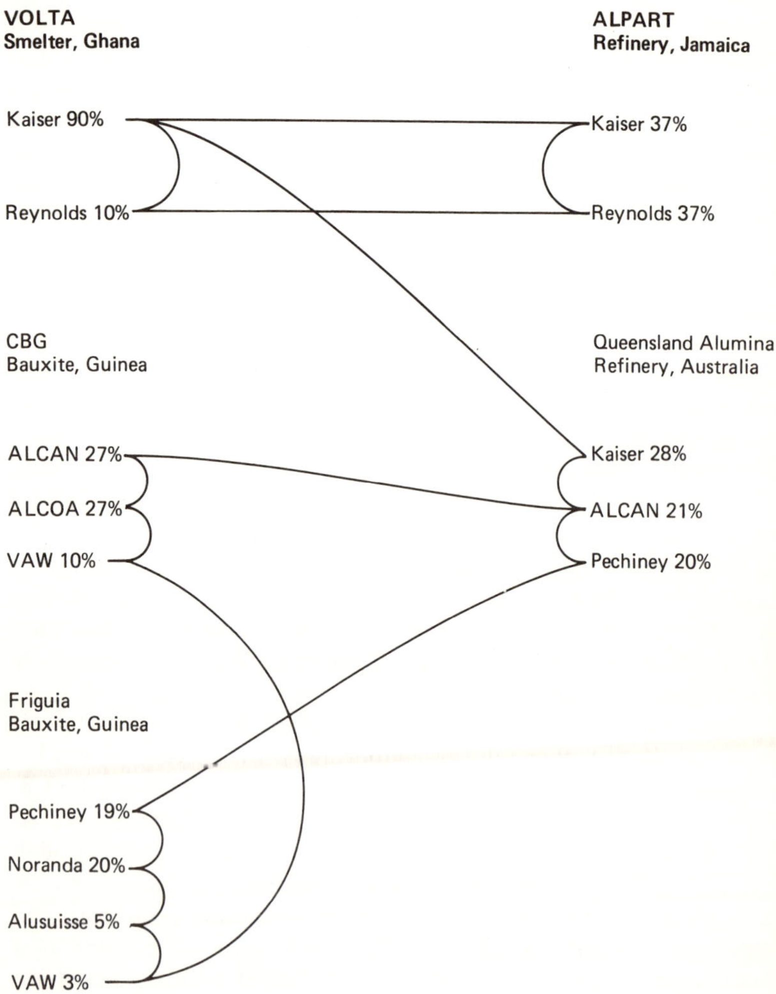

founded in 1972 and headquartered in London. The Institute links aluminium companies from twenty-four countries and promotes industrial public relations, making the public aware of the usefulness of aluminium. It also publishes monthly statistics on production and inventories which could be important for market management activity.

Finally, there is strong evidence that, at least in Europe, 'the club' is still a major behind-the-scenes force. In September of 1978 the European Economic Commission released a 196-page report on its three-year investigation of what is called 'a broadly based cartel'.[8] One of the functions of the old Aluminium Alliance was to prevent low-priced Russian aluminium from depressing world prices by buying up any surpluses from that source. According to the EEC report, in 1963 West European producers created a Liechtensteinian company strangely entitled the 'International Fair Trade Practice Rules Administration'. The function of this association was to maintain a 'fair' price by buying up any surplus aluminium Comecon might wish to get ride of at bargain prices and then reselling it at the going West European price. The investigation also turned up evidence of so-called 'swap deals' formally a common practice in the oil industry. Rather than ship European metal to California or Gulf Coast metal to Europe, an 'old-boy network' allowed the companies to arrange local inter-company swaps to supply local markets, but at the same price, thus pocketing the savings in transportation costs.

Theoretically the club is now defunct. When the EEC announced it was beginning the investigation in 1975 the COMECON producers abruptly dropped the arrangement. None the less, the Commission is still suspicious: 'there is also a feeling among officials that the deal is not ended, merely dormant, ready to emerge again once their backs are turned.'[9]

SUMMARY

The pattern of industry co-operation and conflict is depicted in Figure 3.3. Conflictual behaviour took the form of invading the rival firm's domestic market sometimes by dumping excess production and by aggressive acquisition programmes aimed at threatening the rival firm's independence of action. In other words, the conflictive repertory included various forms of market share competition or 'buyout' attempts. On the other hand, co-operative behaviour involved various

Figure 3.3 Chronology of aluminium cartels (horizontal axis is degree of co-operation)

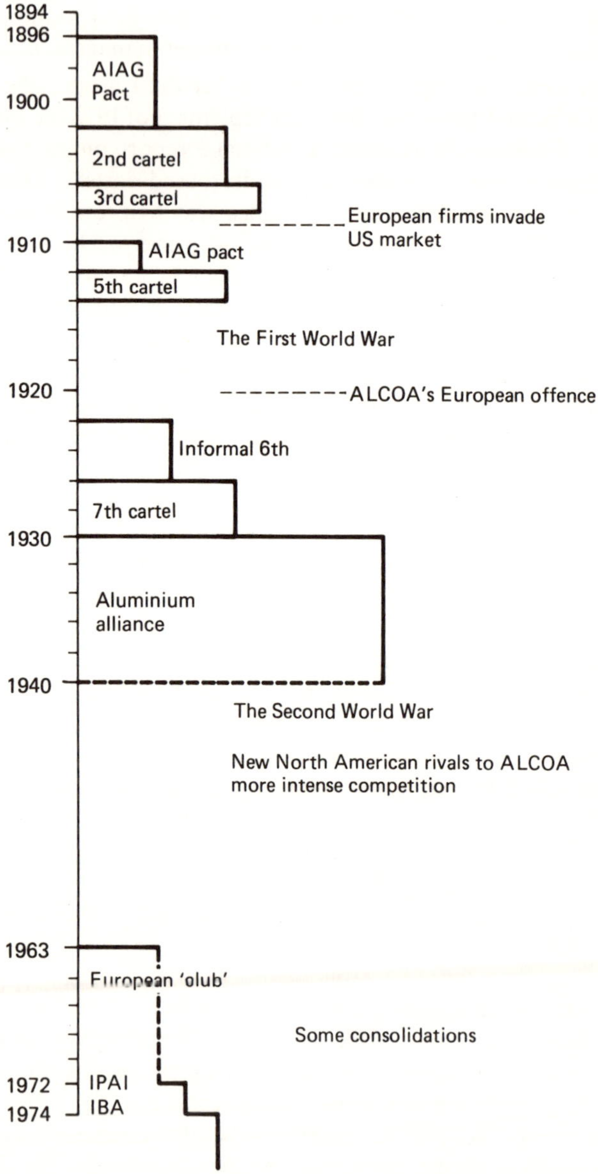

formal and informal agreements to limit competition by controlling prices and dividing up and fixing market shares. The striking feature through this period is the steady perfection of the cartel system of co-operation. Generally speaking, movement from the first to eighth cartel represents a progression to higher forms of collusive behaviour.

Clearly, any theoretical explanation for the aluminium firms' behaviour must encompass both conflictive and co-operative behaviour. Such a model should specify the underlying variables controlling shifts from co-operative to conflictive phases. Two such variables would be change in the world economy and change in technology.

The world economic depression in 1930 is a key variable for explaining the upgrading of co-operation that occurred with the creation of the Aluminium Alliance. Similarly the patent battles of the late nineteenth century explain the lack of co-operation prior to the first agreement. But clearly the main factor for explaining the breakdown of a regime is war between nation-states. War is significant for a number of reasons. First, as a strategic material, aluminium is in great demand. Hence production capacities are expanded rapidly. Likewise, both world wars have brought major technological innovations in the uses of aluminium, particularly in the aviation field. Thus war led to major changes in the relative sizes of the firms and stimulated a period of intense competition for the new markets; this was so after both the First and Second World Wars.

And a second way in which war affects the cartel regime is that nation-states intervened more directly into the industry. Aluminium production became too important to the war effort to be ignored by government. At such times, government to company interaction seemed to take precedence over interaction between the firms. Furthermore, collaboration at the international level may lead to accusations of treason in the firm's home country.

For these reasons, the Second World War ended the Aluminium Alliance, the highest achievement of the MNC's co-operation. The post-Second World War period has been somewhat more ambiguous for a number of reasons. First, ALCOA lost its vehicle for co-operation in the cartel system with Judge Knox's decision in 1950. Furthermore, ALCOA lost its long-held monopoly over US aluminium production, and had to learn to contend with its new domestic rivals. Also, anti-trust sentiment in Europe had begun to develop making open formal agreements much more difficult.

The firms' sensitivity to anti-trust scrutiny has lead them to cover their tracks very carefully. In 1981, the International Bauxite Association (IBA) held a meeting of bauxite producers and aluminium multinationals. In reading the transcript of the *public* sessions, I am impressed by the number of occasions in which a statement by a company executive is followed by a brief statement by the company's lawyer to disavow or 'clarify' statements that may have had collusive overtones. In Chapters 5 and 6 I will show that this instance may well be an example of where there's smoke, there's also fire. None the less, companies are much more careful now than they were in the glory days of Aluminium Alliance.

On the basis of the evidence reported in this chapter, I have suggested in Figure 3.3 that the post-war period can be divided roughly in two. The 1950s and 1960s represent a period of renewed conflict and competition as the market sorted itself out after the Second World War and adjusted to new majors (Reynolds, Kaiser, ALCAN). By the late 1960s and early 1970s, however, after some important mergers, a new phase of co-operation appears to have developed with the creation of the International Fair Trade Practice Rules Administration, IPAI, and the International Bauxite Association (IBA).

This chapter has been a study in oligopolistic behaviour. The material confirms the conclusion of Chapter 2, namely that the Aluminium industry can be characterised as belonging in row three of my Table 1.1 model – demand inelastic and oligopoly. Conditions of oligopoly have created a pattern of behaviour which cannot be simplistically labelled either conspiratorial or competitive. In Marxist terminology, neither Lenin's or Kautsky's predictions are correct all of the time.[10] Under the proper conditions, cartels of increasing sophistication have been formed – at least in the first half of the century. The following two chapters examine another source of potential cartel power – the nations which provide the raw material of the bauxite producers.

4 The Third World Bauxite Producers

The previous two chapters described the aluminium industry and the major aluminium multinationals. Those chapters traced the history of business cartels and portrayed the industry in the 1970s as a classical oligopoly composed of six majors. In this chapter, the focus is on describing the other set of actors important to the bauxite cartel drama to be described in Chapters 5 and 6: the countries where bauxite is mined.

Guyana, Ghana, Guinea and Jamaica were the major Third World producers as of 1974. As such they led the struggle for a New International Economic Order (NIEO) in aluminium. In particular, this chapter traces the quest for consolidation of the right of these producers to control their own natural resources. The participation and nationalisation battles of the bauxite producers placed them in the forefront of Third World nations struggling in the 1970s for a NIEO.

This chapter also acknowledges a major weakness in the bargaining positions of bauxite producers. As Chapter 2 pointed out, aluminium production is segmented into several stages: mining, refining and smelting. It was also noted that processing halves the volume of the product at each stage. Therefore, strong economic arguments exist for locating refineries close to mines and thus reducing transportation costs. Where nearby cheap hydroelectric power exists (as in Guyana and Ghana), economic logic would also seem to favour locating smelters near the mines. Yet, with the exception of Surinam, the major Third World producers do not have integrated aluminium production facilities. As will be seen in this chapter, the Majors appear to follow a policy of geographical diversification of stages of production. This policy greatly increases the costs to the bauxite producer if it contemplates nationalisation. A complete nationalisation of its bauxite mines and complete break with the MNCs would only serve to leave the government with unprocessed bauxite and no market, with profound financial and labour implications. The consequences of this type of action were most evident in the Guyanese nationalisation of 1971.

GUYANA

Both the British and Dutch Guianas became important bauxite producers early in the twentieth century. As ALCOA depleted its domestic sources, it increased its bauxite holdings in Guyana, forming the Demerara Bauxite Co. (DEMBA), and began acquisition of bauxite land in Surinam during the First World War.[1] By 1930 Surinam held 16.2 per cent and Guyana 7.5 per cent of world production.[2] In 1928, the American monopoly spun off DEMBA to its Canadian 'partner' as part of its anti-trust reorganisation (see previous chapter). Thus began the relationship between ALCAN and Guyana which was to culminate in the industry's most bitter nationalisation battle.

Bauxite production rose steadily through the Second World War, and by 1950 Surinam and Guyana together provided 45.1 per cent of world bauxite production. As the strategic uses of the metal increased in the Second World War, so did the strategic importance of these two sources. ALCOA had retained direct control over the Surinam ore, which supplied 65 per cent of the bauxite needed for US wartime aluminium production. Hence, with the fall of the Netherlands to Hitler's Germany, Cordell Hull asked the Dutch government in exile for permission to spend 1000 troops into Surinam. The US force arrived in November 1941 and stayed for the duration of the war.[3]

The situation in Guyana was complicated by economic rivalry between Britain and the US. During the First World War, the Colonial Office had expressed some concern about American control of a mineral so important to the Empire, but with the combination of US government pressure, the discovery of African bauxite sources, and the lengthening of the distance between ALCOA and ALCAN (particularly after Judge Knox's 1950 ruling), British concern over US domination declined. After the 1950 formal separation, ALCAN became critically dependent on bauxite from 'its' Guyanan monopoly.

Yet most of the benefits of this 'mutually dependent' relationship flowed out to the global company. DEMBA took on the characteristics of a classical company town and an outwardly focused foreign enclave. ALCAN paid no royalties on the bauxite mined on its own land. On Crown land, it was charged 10 cents a tonne, but in the absence of an open market for bauxite, the Colonial Office was dependent on ALCAN's own assessment as to the value of the ore.

In the Colonial Office's eyes, its bargaining position was further eroded by political developments in the country. In 1953 Cheddi Jagan won his first electoral victory, prompting a British reoccupation four months later to halt the 'radical drift' of the country. At the time the policy of both Cheddi Jagan and Forbes Burnham towards the company was nothing more radical than to suggest it should pay more taxes.[4] But the British Governor believed that Guyana was then seen as a poor political risk to foreign investment. According to a well-known business investment pronouncement, profit must be inversely related to level of risk. Therefore, when Reynolds entered the country in 1953 through a small company buy-out it was considered only appropriate that its faith in Guyana's future be rewarded with a ten-year tax moratorium. Graham claims that between 1917 and 1969, only G$21 million worth of government revenue[5] was received in return for G$1200 million worth of bauxite.

Jagan returned to office in 1957 and survived to 1963, despite British and American destabilisation efforts. At that time, Burnham's government dropped a back taxes complaint against ALCAN, which reciprocated by promising (but never building) a smelter to complement a small alumina plant built previously.

The continuing failure of the foreign aluminium giants to provide more visible benefits to the Guyanan economy toughened even the Burnham government's resolve and rhetoric by the late 1960s. Bauxite revenues were still considered exploitatively low. Compared to Jamaica with its new refinery and Surinam with its new smelter, Guyana appeared to be falling behind in value added production. By 1970 Guyana had begun to sound out Jamaica and Surinam on the possibility of forming a common bauxite council to confront the aluminium firms. The then ruling Jamaican Labour Party showed no interest in this Guyanese overture, in part because it was concluding negotiations on its own for a small increase in Jamaican bauxite tax. The Guyanese leaders also attempted to pressure the firms into making greater use of local intermediates, such as local flour (starch) in the alumina process.

By the end of 1970, the Guyanese government had chosen majority participation as the only solution. Prime Minister Burnham was at first reluctant to go so far as nationalisation, but the failure of the equity negotiations forced the move. As one leading business publication put it in 1971, 'The Prime Minister may have been a victim of his own hand. The government had to pass up less drastic measures to accommodate the nationalist fervor it had whipped up since Guyana became a socialist, cooperative republic early last year'.[6]

Graham is correct in calling the Guyanese nationalisation 'the start of a new wave of economic nationalism in the Caribbean which came to a climax with the formation of the International Bauxite Association in 1974'.[7] However, his claim that an aluminium multinational was 'successfully challenged without any undue consequences'[8] is overstated at best.

The former ALCAN subsidiary, renamed Guybau, immediately faced a series of grave problems; in particular, the flight of the entire North American management team, and the lack of a well-developed marketing system for its bauxite and alumina. In compensation, Guyana agreed to pay $53.5 million for the facilities, a price slightly higher than the company's earlier estimates of value. 'The Guyanese government is obviously trying to keep its relationship with ALCAN as cordial as possible'.[9] The government still needed ALCAN's marketing outlets and ALCAN agreed to buy half of Guybau's output. The government report on Guybau[10] glosses over many of the management and marketing problems which necessitated the closing of the alumina plant for a brief period, and bauxite production dropped slightly as Table 4.1 shows.

Table 4.1 Bauxite production in 1000 metric tonnes[11]

	1969	*1970*	*1971*	*1972*	*1973*	*1974*
Guyana	4 306	4 417	4 234	3 668	3 621	3 049
Jamaica	10 498	12 106	12 543	12 989	13 601	13 328

The increase in Jamaican production over the same period suggests that the North American firms may have made up the difference from that source. But the situation in Guyana seems to have taken a profitable turn by 1976, thanks perhaps in part to the cartel. 'Its 1975 after-tax earnings tallied perhaps G$25.2 million, up from G$19 million in 1974 and G$8 million in 1973. Since nationalisation, the firm has seen its total capital more than double, to G$227.6 million; the return on total mean capital is 13.4 per cent'.[12] Guyana also had ambitious plans to tap its hydroelectric potential and build an aluminium smelter.

The ALCAN nationalisation put Reynolds on notice that sooner or later the Guyanese government would make similar demands of it. In 1974, as part of the IBA offensive, the government acquired 100 per cent control of Reynold's mines as well.

GHANA AND GUINEA

In part, the British government's resistance to ALCOA's move on Guyana was broken by the discovery of other bauxite sources in the Commonwealth. For example, bauxite was discovered in the Gold Coast (Ghana) in 1914. However, the preference of the British Aluminium Company (BAC) and the Colonial Office was that these Ghanaian sources not be developed but rather be held as a strategic reserve. The BAC was quite content to continue its traditional reliance on the large French domestic bauxite mines. However, the fall of France and the end of French exports (along with sufficient inducements by the Colonial Office), led to BAC mining in 1941 at Awaso. After the war, BAC first considered closing the mine, then kept it open in spite of relatively high costs. In part, BAC's decision to stay was based on the desire to have a hand in any potential smelter scheme.

Given Ghana's bauxite and obvious hydroelectric potential, an integrated aluminium industry seemed a logical development to many entrepreneurs. Graham's 1982 case study of Ghana's Volta River Project documents numerous plans and efforts along this line. From 1951 on, sporadic negotiations between Ghana, ALCAN and sometimes BAC took place in London for a dam and an integrated aluminium industry. But these negotiations had completely broken down by the occasion of Ghana's independence in 1957. Given that both ALCAN and BAC had sufficient smelting capacity elsewhere, it is easy to suspect that the companies' motives were obstructionist and pre-emptive. After independence, Nkrumah turned to the two new, hungry American companies: Kaiser and Reynolds. Kaiser in particular needed smelting capacity and so took the lead in the negotiation of the formation of the Volta Aluminum Company (VALCO). Eventually, Kaiser took the lion's share of ownership (90 per cent to Reynolds' 10 per cent).

While Nkrumah hoped the Volta Dam would spark Ghanaian industrialisation, the World Bank and the Kennedy administration believed that the project would require a major customer to be commercially viable. Hence, they made financing for the dam contingent on a smelter agreement. In 1962 Nkrumah signed the Master Agreement by which Ghana finally got the Volta dam, power plants and national grid but only by locking the country into a 30 year agreement to sell much of its power to the VALCO smelter at a rate that makes it one of the world's lowest cost operations.[13] The dam was finished in January of 1966 one month before the first military coup. The following year the smelter went into operation.

Nkrumah (and subsequent Ghanaian governments) had hoped that Kaiser and Reynolds would integrate backward into bauxite mining. Indeed, Reynolds gained control over BAC's Ghanaian mines in its 1957 acquisition of BAC. Now only one stage was missing; but no alumina refinery has been built. Instead, Reynolds ships the bauxite out to the BAC's traditional refiners and ships its own alumina in. In the early 1970s, the BASCOL consortium (Kaiser, Reynolds and several Japanese companies) negotiated over the possibility of further bauxite mines and an alumina refinery. In 1975 high energy prices drove the Japanese out and the consortium collapsed.

Today, Ghana continues to be a minor, 'reserve' bauxite producer. Awaso is still run by the BAC (owned by Reynolds) and its production peaked in 1972 at 350 000 tonnes compared to neighbouring Guinea's 1972 production of 2 600 000 tonnes. As will be shown, the 1974 IBA initiative resulted in the partial nationalisation of the Awaso mine under the new name of the Ghana National Bauxite Company.

Kaiser and Reynolds may feel that providing Ghana with an alumina refinery will greatly enhance the threat of nationalisation of all their interests. But from Ghana's point of view, the aluminium companies are playing the classic multinational game of maintaining dependence by locating different stages of production in different countries.

Guinea, in contrast to Ghana, has extensive bauxite mines and reserves and a refinery to make alumina, yet it has no aluminium smelter despite enormous hydroelectric potential. As part of the French empire, it is no surprise that historically the development of Guinea's large bauxite reserves only began after the Second World War upon the depletion of France's domestic bauxite mines. Today Guinea supplies 75 per cent of French needs[14] and ranks second in world bauxite production. Guinea is also important to Canada: with ALCAN's 1971 departure from Guyana, Guinea became ALCAN's main source of bauxite. The country's two main bauxite fields are Fria and Sangaredi.

In 1958, the French colonial government negotiated a 75-year mining agreement with the Fria Consortium (Alusuisse, Pechiney, VAW, the BAC and Noranda, a Canadian Company). The agreement granted the consortium concessions on import tariffs and taxes and provided no option for even partial government or local participation in ownership. Mining began in 1960.

ALCAN first expressed interest in developing the bauxite potential of the Sangaredi Plateau in the 1950s. But upon Guinean independence in

1958, the new government demanded partial ownership and ALCAN quickly departed. The government also expressed its dissatisfaction with the FRIA arrangement, but in 1958 the idea of government joint-ventures was unacceptable to the Majors. Yet, in a pattern similar to that found in the oil industry, Guinea accomplished a breakthrough agreement with a 'fringe' company, Harvey Aluminum of the US.

Harvey's struggle with the Majors was described in the previous chapter. Lacking alternative sources of bauxite, Harvey was prepared in 1963 to pick up the former ALCAN project with 49 per cent government ownership and total government control of the railway and port facilities. The infrastructural requirements lead the government to the World Bank and became a major hitch in the deal for Harvey. After a $2 million technical study, the Bank ruled that Harvey's proposed production rate of one million tonnes per year was too small to justify such a large loan and that a six-fold increase in production was required. This demand put the project well beyond Harvey's resources and forced the company to seek partners. Eventually a consortium was arranged with a signed 20-year agreement in 1968 with the government and the World Bank. Guinea got its loan and kept its 49 per cent share of the mine. The Halco Mining consortium controlled the remaining 51 per cent, shared out 27 per cent to ALCOA, 27 per cent to ALCAN, 20 per cent to Harvey, 10 per cent to Pechiney, 10 per cent to VAW, and 6 per cent to Montecatini (an Italian company).[15] Thus, on the one hand the Majors had been forced to accept partial ownership by a Third World government. But on the other hand, they had in turn forced rival Harvey to surrender exclusive access to one of the world's largest bauxite mines. Production began in 1973 and grew by 1979 to an annual capacity rating of 9 million tonnes (though Harvey has since disappeared).

Fortified with this success, the Guinean government was able in 1973 to renegotiate the FRIA agreement, giving it 49 per cent ownership and 65 per cent of the profits. In 1974, a third bauxite mine came into production at Kindia, built with Soviet technical help and 100 per cent government owned. By long-term agreement, 50 per cent of its new output goes to pay for the Soviet assistance while 40 per cent pays for purchases of Soviet goods, leaving 10 per cent to the government.[16]

The government's attempts to develop an integrated industry have met with considerably less success. The FRIA consortium processes its bauxite into alumina locally and then exports it to company smelters elsewhere (Pechiney in Cameroons, Reynolds (the BAC) in Ghana). As with Ghana, the Majors consider Guinea 'politically risky' and hence have followed a diversification strategy in terms of processing. Indeed, for a period in the early 1960s, all three countries so far discussed appeared politically unappetising to the North American Majors for further investments.

JAMAICA

By contrast to Guyana, Ghana and Guinea, Jamaica's political acceptability in the 1960s, large reserves and closeness to Gulf coast smelters seemed to make it the 'jewel in the crown' of the North American industry. Because of its crucial importance to the industry and its leadership role in the International Bauxite Association under Michael Manley Jamaica is covered in this study in more detail than the other major bauxite countries.

Jamaica can be described historically as a modified plantation-type developing economy.[17] From the seventeenth to early nineteenth century, sugar grown on plantations with African slave labour was the dominant export, followed later by bananas. The abolition of slavery in 1834 established a pattern of surplus labour supply which has plagued the economy down to the present. Unemployment problems have generated social frustration leading to periodic strikes and riots such as those of the 1930s. The agricultural sector is still characterised by a dualism between large plantations and small plots. In 1943, 1.4 per cent of the nation's farms owned 66 per cent of the land and 92 per cent had only 21 per cent of the land. Hence, Jamaica fits the Latin American pattern of skewed land distribution with land reform as a main national priority.

Jamaica's foreign investment policy had followed the lines of an open, capitalist developing economy. The aluminium MNC invasion beginning in the 1950s represented the single largest inflow of capital: $145 million in the years 1950 to 1966. This inflow has had a dramatic impact on aggregrate measures of growth: annual growth rates of GDP increased from 0 per cent in 1950 to 10 per cent in 1967[18] and 12 per cent by 1972.[19] The MNC investments were the single largest source of tax revenue, contributing 15 per cent of government expenditures in 1967. On paper, the rapid expansion of the bauxite sector made Jamaica one of the fastest growing LDCs before 1974.

Nearly all of Jamaica's bauxite goes to the US through intra-company transfers by the three major US firms. The only exception is ALCAN's alumina refinery which supplies its smelters in Canada and Scandanavia. In 1969 a consortium of ALCOA, Reynolds and Kaiser built an alumina refinery, Alpart, to export alumina to the US, but the major export is still raw bauxite.

Kaiser, Reynolds, and ALCAN entered Jamaica in the early 1950s. In part, this move was directly sponsored by the US government. The first source of the US involvement grew out of the government's desire to create fully integrated rivals to ALCOA. While the Surplus Properties Administrator provided Kaiser and Reynolds with alumina refiners and aluminium smelters, it could do little in the way of supplying bauxite sources. Reynolds had a few Arkansas properties of low quality, but ALCOA and ALCAN dominated the well-known foreign sources in the Guianas. Hence the discovery of cheap Jamaican bauxite was a godsend for Reynolds and Kaiser. Secondly, the US Government favoured Jamaican bauxite for national security reasons. Given the Korean war and the general cold war context, the government regarded Jamaica as an easily defensible source of the strategic material. Hence the government extended Reynolds and ALCAN aid credits to establish mining operations there.

Many additional factors have contributed to make Jamaica attractive to the North American MNCs. First, Jamaican bauxite had lower extraction and transportation costs than other sources. The mining overburden (the rubble which must be cleared to extract the ore) is very light in Jamaica. Bauxite sites are more accessible than the already heavily worked Guyanan deposits and its costs less to transport bauxite to the US from Jamaica than from South America. This cost factor led Kaiser and Reynolds to convert their nearby Gulf Coast facilities specifically to the processing of Jamaican ores. Bauxite from Surinam which is 'rich' in alumina is also 'rich' in noxious impurities such as silica. A second attraction is the huge size of the Jamaican reserves, allowing for the long-term commitment of investments. And finally, there was the relative 'political safety' of Jamaican investment. After independence, both political parties – the Jamaican Labour Party and the People's National Party – were anxious to reassure the companies. The alternatives of Guyana and elsewhere in the period 1957–64 were none too appetising for the firms. The 'Marxist' government of Dr Cheddi Jagan was in power in Guyana. While ALCAN expanded its operations there, the more suspicious Reynolds waited until the anti-Communist coalition came to power in 1964 before agreeing to any further investments. Although ALCOA had a strong base in Surinam, it used a newly discovered source in the Dominican Republic for its incremental expansion. But the withdrawal of US support for Trujillo in the late 1950s left ALCOA in a tenuous position, and soon thereafter ALCOA entered Jamaica also.

Decisions to expand mining capacity in Jamaica have been dependent on corporate strategies and the expansion of aluminium capacity by the North American industry. Both factors lie in the hands of the aluminium MNCs; the Jamaican government could only offer incentives. In particular, the government has encouraged alumina refining to increase value added on the island. ALCAN built two alumina processors: 'Kirkville' with Marshal Plan funds in 1959 and Ewarton in 1960. Alpart represents a similar move on the part of the Americans. In the late 1960s Revere entered and discussed alumina facilities. But the firms already had technical incentives to locate alumina processors in Jamaica given the relative costs of shipping alumina as opposed to bauxite.

By the early 1970s, all firms leased extensive tracts of land in Jamaica and invested to a small extent in agriculture to meet government land restoration and idle farmland legislation. Given the persistent land inequalities and agricultural needs of the island's large population, the government's past policies of liberal land leases drew increasing attacks from many quarters. In addition of the MNCs had established its own ports and storage facilities (such as 'Kaiser' and 'Port Kaiser') and all but ALCAN had their own transportation facilities.

By the early 1970s the Jamaican government, and in particular the PNP government of Prime Minister Manley, had grown increasingly sensitive to both the relative share of benefits the bauxite industry distributed locally and the overall impact on the local economy. The economic policy of the 1950s and 1960s had been premised on the liberal model of development which portrays direct investment as an 'engine of development' and creates a 'leading sector' which pulls up production in the rest of the economy through the use of local intermediate goods. But while gross level indicators continued to rise, by the late 1960s the 'spillover' effect still seemed quite insignificant. Norman Girvan (1971) calculates that the local share of the value of output in the Jamaican bauxite industry has fluctuated from 33 per cent to 51 per cent as depicted in Figure 4.1. Furthermore, he finds that payments for materials, wages, and salaries make up an insignificant part of the local share, with tax payments comprising the greater proportion. In part, this is due to the absence of many intermediates in mining processes in general. But the MNC overseas parent makes the crucial decisions on factor proportions and local inputs and outputs. As a result, the parent has chosen to use mostly non-local inputs, many of which the vertical combine already produces for itself or buys in volume from US domestic producers, which it can ship cheaply to Jamaica in its returning empty ore ships.

The low local share is also a result of the higher level of capital intensity chosen in bauxite operations and the fact that, since the capital is foreign owned, the higher returns on capital flow into foreign hands. The bauxite industry in Jamaica is very capital intensive: Girvan

Figure 4.1 Export value and current local payments of the bauxite industry

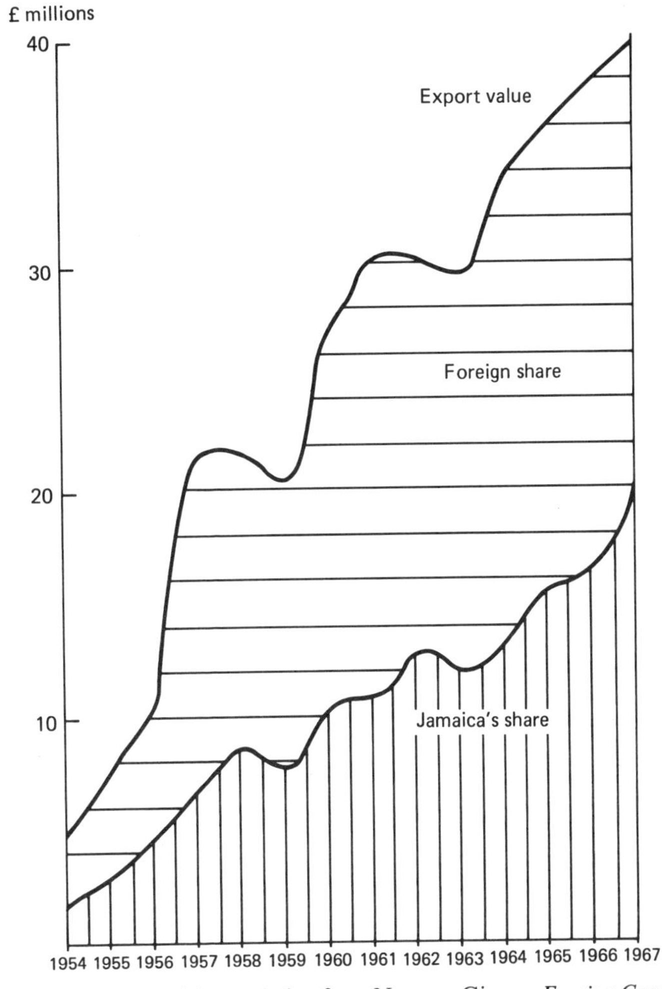

Source: Reproduced with permission from Norman Girvan, *Foreign Capital and Economic Underdevelopment in Jamaica*, Institute of Social and Economic Research, University of the West Indies, 1971, p. 40.

calculates over $12 000 in assets per worker as opposed to a general manufacturing range of $400 to $2 000 per worker. Each company had a wide range of latitude in determining the capital to labour ratio in its mining operations (essentially an earth-moving operation). In general, they chose capital intensive production methods because they were more appropriate given the requirements of a constant flow of bauxite to alumina plants. However, a capital-intensive method is certainly inappropriate from Jamaica's perspective, given the island's overabundance of labour.

The result is that the MNCs' bauxite operations comprise a classic self-contained autonomous sector in the Jamaican economy. The government's response has been to pressure the firms to set up alumina production on the island, thus increasing value added, stimulating the production of these inputs and increasing tax revenues. But the existing alumina refineries import most of their intermediate goods. The companies have largely ignored attempts by local firms to develop local production of intermediate goods such as caustic acid and chlorine.

Another concern of the government is that the foreign bauxite sector may be introducing distortions into the economy of the island. Aside from government tax revenue, the only other local incomes of any consequence are payments to local transportation and construction industries. The three American companies transport most of their bauxite to port themselves, but ALCAN transfers its alumina to port via the Jamaican Railway Corporation (JRC). ALCAN is thus a major customer of the JRC for which it receives a reduced shipping rate. The remainder of local income is received by the construction industry and fluctuates on the basis of investment expansion by the firms. New investments by the companies represent huge construction projects for the local industry a temporary, dominant share of its business. Given the cyclical nature of investment patterns, however, Girvan shows that this only serves to exacerbate the unemployment problem since its impact on the lives of construction workers is volatile, periodically throwing thousands out of work. Again this distortion is due to a problem of relative size – the small size of the construction industry by contrast to the huge size of the firms' projects – and the lack of alternative big projects in the developing country to take up the slack.

The companies' wage payments, while modest in total, have a dramatic impact on the labour situation in such a small developing economy. As Mitchell (1968) points out, the companies operate on the basis of American labour practices and wages. In 1951 the United Steelworkers of America helped the local National Workers Union to organise the bauxite workers. In this they were assisted by the firms, who preferred the American shop-steward system of processing grievances. With time, the bauxite workers have become the core of the National Workers Union.

> the bauxite companies established wages which were often beyond the capacity of other employers to match, with the result that they recruited heavily from other industries. Due to the high degree of mechanization, the numbers they employed were relatively low. A highly paid elite group of workers emerged which may have been a mixed blessing for the island.[20]

These labour policies and the general insularity of the MNCs have ironically eroded the support of a group that could have potentially supported their interests – the local business élite. Kuper (1976) reports little support from such indigenous leaders for the bauxite firms:

> Jamaican businessmen are particularly well-organised to press their demands, but their interests are not always congruent with those of the most powerful economic bodies, the multinational corporations. They, with their monopolistic positions, are most concerned with predictability, with maintaining control of the long-term development of their operations . . . The bauxite companies have created new and formerly undreamt-of aspirations for the workers. They have also worked closely with the unions, and greatly strengthened them in consequence . . . The distinctness of these economic interests is reflected in the generally contented attitude adopted by Jamaican capitalists when, recently, the PNP government turned the screws on the bauxite companies.[21]

This 'turning of the screws' can be seen as the direct result of the failure of past government policies either to raise the local share of total output to change corporate decisions perceived as having a distorting effect on the local economy. With the formation of IBA (the bauxite cartel) the government attitude toward the companies took yet another twist. The cartel created a new order that gave government and MNCs a shared interest in high aluminium prices.

This section has traced the evolution of a Third World government's attitudes toward the aluminium MNC from total acceptance of the neo-classical or liberal model of development to a position more aligned with the New International Economic Order (NIEO). The promised benefits of foreign investment through the 1960s failed to manifest themselves; rather indications of economic distortions multiplied. The end result was to lead the government into a tougher bargaining position with demands for renegotiation of contracts, high taxes on bauxite, and government participation in the control of local subsidiaries.

What factors can explain the shifts in attitude by the government? It appears that a learning process is at work not unlike the perceptual process described by Moran (1974) for Chile. Internally the government became disillusioned with the neo-classical model at the same time that new perceptual models became available externally through the development of the NIEO in the setting of the UN system. Furthermore, the companies may have been slow in seeking local coalition partners (such as indigenous businessmen) perhaps due to their inexperience with the Jamaican political system.

CONCLUSION

By 1974 across Third World bauxite producers a common pattern of grievance had emerged which became intermeshed with the general call for a New International Economic Order. In all countries, in spite of large-scale foreign investment, the bauxite industry failed to be an 'engine for development'. The Majors operated as economic enclaves linked to global processing systems with virtually no local purchases or spillovers.

The NIEO proclaimed the rights of Third World states to control their own natural resources. By 1974 in the bauxite industry the struggle for recognition of those rights was well underway. First, Guinea, using a strategy common to oil producers, had by 1963 gained the right to government minority participation in a new bauxite project from a minor company desperate for bauxite and willing to break the existing 'rules of the game'. By the subsequent 1968 agreement with the HALCO consortium, three Majors – ALCOA, ALCAN, and Pechiney – came to accept this principle. Then, in 1971, the Guyanese nationalisation pitted a producer against a long-entrenched company (ALCAN) with a near monopsony position. The importance of that source to ALCAN and the company's seeming upper-hand may well have lead to the company's bargaining intransigence. The resulting total nationalisation of its assets marked an important stage in the struggle for producer control. None the less, this dramatic break with the Majors was not without negative consequences for Guyana (production levels, technology and forward integration) and may have served as an important lesson for the Manley government in Jamaica. In 1973, Guinea established 49 per cent government control of an existing bauxite operation.

In 1974, as part of its general offensive, Jamaica steered a path between the Guyanan and Guinean models, demanding control but leaving the Majors with a minority interest in the operations. Jamaica was supported in its actions by the collective militancy of the IBA. The same year, Guyana nationalised Reynolds and Ghana took a 55 per cent share of its own local bauxite mine. The story of the Jamaican nationalisation and its consequences is told in Chapter 6.

A second major principle of the NIEO is the right of the Third World to a fair return on raw materials. Both Guyana and Guinea struggled with tax concessions granted to the MNCs by colonial governments. All bauxite producers faced a common lack of independent information on the value of the bauxite and hence lacked informed assessments of fair return. By 1974, it was clear that only collective action would earn the producing countries a greater return on their bauxite.

5 The Impact of the Bauxite Levy on Company Profits

Chapter 4 demonstrated how the calls for a NIEO resulted in demands for recognition of rights of ownership among the bauxite producers. This objective was usually obtained by either full or partial nationalisation of the mining process. The present chapter traces the bauxite producers' success on a second NIEO front – the demand for a greater return on the bauxite itself. The chapter concludes with an analysis of the impact of the bauxite producers' price offensive on the MNCs using a quasi-experimental design to track the changes in their profits. Evidence suggests that by 1977–8 the aluminium multinationals might have found co-operation with the bauxite cartel 'not unprofitable'.

THE FORMATION OF THE INTERNATIONAL BAUXITE ASSOCIATION

As early as 1972 Michael Manley may have been contemplating action to raise the price of bauxite to the industry. Speaking before the UN General Assembly in October, he stated:

> The time has come to reconstruct the basis on which the gains arising from the exploitation of resources are shared between those . . . who provide the capital and technical know-how for the process of exploitation, and those . . . who own the resources and provide the essential infra-structure and labor force.[1]

Throughout 1973, Manley and Prime Minister Burnham of Guyana promoted the idea of an association of bauxite producers publicly and privately. Their organising efforts led to a first meeting on 19 November 1973, of six major bauxite nations (Australia, Guinea, Guyana, Jamaica, Surinam and Yugoslavia) in Belgrade to form 'an OPEC-like group to control world bauxite prices'.[2] The actual founding of the International Bauxite Association (IBA) then followed at a March 1974 meeting in Conakry, Guinea. Joined now by Sierra Leone, the founding seven members stated as one of the association's main goals – 'to secure for member countries fair and reasonable returns from the exploitation, processing, and marketing of bauxite'.[3]

Undoubtedly the major triggering mechanism for the formation of the cartel was the success of OPEC oil price increases both as providing a positive model and as generating a crisis in the balance of trade situation of most of the IBA members. Jamaica, for example, had its oil import bill triple between 1972 and 1974. Like other oil importers, Jamaica found itself compelled to either increase the volume of its exports (unlikely) or increase the value of them. Bauxite became the obvious target.

In March of 1974 just after the Conakry Conference, Jamaica announced the imposition of a substantial new 'production levy' to boost the share of total output which Jamaican bauxite captured. In the past the bauxite nations had been stymied in seeking higher returns for the raw material by the absence of either an open bauxite market or bauxite price information. Jamaica, for example, had to rely on the companies' estimates of the value of the bauxite for tax purposes. But in setting its new production levy, Jamaica solved this problem by linking the tax to the ingot price of aluminium and demanding participation (but not nationalisation) in the local mining operations, so as to acquire the needed internal financial information. Under the new law the levy was set at 7.5 per cent of the quoted price of a finished aluminium ingot. While still contributing a relatively modest share of the cost of aluminium, the levy raised tax revenue for Jamaica seven-fold.

Jamaica's gamble on unilateral action appeared justified by the summer of 1974 when the other Caribbean producers (Guyana, Surinam, Haiti and the Dominican Republic) followed suit and imposed levies to match the Jamaican tax. Guyana further raised the pressure on the companies by announcing the nationalisation of Reynold's bauxite mine.

PREDICTING COMPANY RESPONSE

At this point the aluminium majors' reaction becomes vital. Concerted action on the part of the companies may have well rolled-back the successes of the bauxite cartel in imposing the levy. One might assume the companies would respond with unmitigated hostility. But the model adopted in Chapter 1 suggests a more circumspect analysis. Under some conditions our model suggested that company profitability may not be damaged by the creation of a producer nation cartel. Indeed, the major findings of Chapter 2 and 3 of this study were that the demand for aluminium was inelastic and the industry structure was essentially that of an oligopoly with a fringe. Referring to Table 1.1, the prediction of the model under these conditions would be for company profits to remain constant or even increase after the formation of a producer cartel. If this is the case, then our model would further predict that the company response would be passive or perhaps even co-operative. The remainder of this chapter is devoted then to adopting a research design to establish what happened to company profits after 1974.

THE QUASI-EXPERIMENTAL RESEARCH DESIGN

The recurrent criticism expressed about attempts to show oil company profiteering after the OPEC price increases of 1973–4 (such as Figure 1.1) is that without a control group there is no way to establish price increases as the source of the increased profits and not some other variable such as a general business upturn in 1973. Campbell and Stanley (1966) suggest the employment of quasi-experimental designs (specifically design number 14) to deal with this sort of problem. Represented in the following manner

O O O OXO O O O

O O O O O O O O

this design involves multiple observations (O) of the experimental group (in this case the company profits of the cartelised industry) before and after the intervention (X, the cartel price increases) as well as concurrent observations of a similar industry which experienced no such intervention. This design benefits from the fact that it allows two sets of comparisons; (a) the experimental industry to itself before and after the intervention, and (b) that industry to the second or 'control' industry.

Concerning issues of validity, Campbell and Stanley argue that Design 14 is relatively free of the internal validity problems (history, maturation, etc.) that plague other designs, but is troubled by issues of external validity or generalisability. Since the testing involves after-the-fact examination of financial reports, I would expect the test itself to be non-reactive (have little impact on the phenomena under study). None the less, the generalisability issue remains. It is interesting in itself to know whether a given industry did realise benefits from cartelisation. But to then suggest any other industry outside the two studied would likewise benefit could not be done without extreme caution.

Having selected the aluminium industry as the experimental group, the task became one of finding another industry with similar characteristics to serve as the control group. Specifically, a variety of other major metals[4] were compared to aluminium on the following criteria:

(a) *Concentration and Characteristics of the Major Exporters* Both bauxite and copper are produced mainly by a small number of LDCs. Iron ore, tugsten, and nickel all have major DC exporters. Tin, however, is produced primarily by LDCs.

(b) *Concentration and Characteristics of the Major MNCs Involved in Extraction* Both bauxite and copper are produced by a handful of large vertically-integrated firms. And both these industries are dominated by North American firms. Iron ore, tungsten, and tin are not so vertically-integrated. Tungsten and tin alike involve many non-North American firms which could lead to data gathering problems.

(c) *Demand Elasticity* Demand for bauxite and copper is considered to be inelastic in the short-to-medium term. Demand for most of the other metals named is also inelastic in all but the very long-run, though tin is considered more elastic than the others.

(d) *Substitution* Bauxite and copper are not near-substitutes for each other. Under some limited conditions, aluminium (the final product of bauxite) and copper can be intersubstituted, but as will be shown those conditions did not apply in the 1970–78 time frame of this study.[5] On this basis copper was chosen as the control industry for the design.[6]

Operationalisation of Profits[7]

In assessing the impact of the cartel increases on the firms, many alternative measures of profit exist which might be employed. I considered four potential indicators: return on book value, earnings per share, the price to earnings ratio, and change in earnings alone. The latter was selected as the operational measure for the reasons given below.

In his recent book on the copper industry, Narvin (1978) makes use of the indicator, return on book value. However, this indicator is seriously flawed for a number of reasons. First, there is the fairly arbitrary nature of book value, a measure easily manipulated for corporate fiscal needs. Book value may be inflated for some financial needs such as expropriation negotiations and devalued for others such as tax assessments. Also book value depends heavily on the particular depreciation method employed by the firm.

Earnings per share and price to earnings ratio are both good comparative investment measures, but tend more to reflect the stock market view of the firm. Earnings per share may depend on the amount of stock outstanding. The two components of the price/earnings ratio (P/E ratio) cause similar problems. A drop in the P/E might indicate a decline in the price of a share of stock or an increase in earnings. Furthermore, financial theory states that if profits rise, the price of the stock is likely to rise in an offsetting manner.

The most commonly accepted measure in business research is the change in earnings alone. For this reason, I have chosen to construct an index of relative percentages of a given base period. In order to generate a sufficient data base, I decided to use after-tax earnings reported on a quarterly basis.

Selection of the Firms

The availability of quarterly information raised a problem for the inclusion of certain firms. I could not find quarterly information for the two large European firms, Pechiney and Alusuisse. A further complication arose for the Pechiney Ugine Kuhlman conglomerate. Its annual report gave a breakdown of 1973 sales:

Aluminium	35.3%
Copper fabrication	22.1%
Mining-Electrometallury	7.9%
Steel and Titanium	7.4%
Nuclear and Special Proj.	7.8%
Chemicals	19.5%

A sense of the overall size of this firm can be gained by considering that with only 35.3 per cent of its activity in aluminium, Pechiney ranks as the fourth largest aluminium firm in the world.

A somewhat similar problem exists for Martin Marietta (Harvey). After 18 April 1972. Harvey's profits are reported as Martin Marietta Aluminum though separate from Martin Marietta. But in 1974 the aluminium firm's earnings disappeared into the parent's earnings. The firm's 1978 annual report shows that aluminium contributed only 33 per cent to earnings before interest and taxes. Similar changes in corporate identification generated problems for the other minor independents. Therefore, I have selected four of the six majors for consideration is this study: ALCOA, ALCAN, Reynolds and Kaiser.

Narvin (1978) lists the six major US copper firms by asset size as

Kennecott, Anaconda, AMAX, Phelps Dodge, ASARCO and Newmont. Anaconda has been removed from my list for several reasons. Anaconda has a moderate interest in aluminium through its subsidiary, Anaconda Aluminum. More importantly, in 1977 it was absorbed by Atlantic Richfield; hence profit data in recent years are unavailable.

Both Anaconda and Kennecott lost assets in Chile, but, as Narvin reports,[8] Anaconda was deriving three-fourths of its profits there, whereas Kennecott earned only 11 per cent from Chile. Furthermore, the take-over occurred in two stages, the first 51 per cent in 1967, well before the beginning of the time-series. The second expropriation in 1971 ironically catapulted Kennecott into the lead as the world's largest copper firm. After carefully comparing Kennecott's earnings to the industry aggregate, I concluded that the impact of Chile's actions on Kennecott appears to be minimal due in part to the small proportion involved, the receipt of US governmental agency insurance, the use of tax losses, and the eventual payment of compensation by the Pinochet regime. The Kennecott nationalisation may be viewed as roughly equivalent to Guyana's nationalisation of Alcan's properties in the same year, 1971.

Calculation of the Index

I collected quarterly reports of after-tax earnings for each of the nine firms from *Moody's Industrial News Reports* for all quarters from the autumn of 1971 to the autumn of 1978 inclusive. The next step was to adjust for the inflation rate. The OECD *Main Economic Indicators* provided quarterly price indices disaggregated by sector using 1970 as the base year. The years after 1976 used 1975 as the base year and therefore had to be translated onto the 1970 base year scale.[9] I selected the price index 'Metals and Metal Products' as appropriate, and adjusted quarterly profits of all firms to reflect 1970 prices. This base year, 1970, was a time of strong profits for both industries due to the high demand for aluminium and copper products during the Vietnam War. Consequently, much of the time-series lines for both groups are below the 100 per cent of base-quarter point. To construct this time-series, I summed the firms' quarterly profits in each industry, and then calculated the relative percentage of the base quarter (the average of 1970s four quarters). This procedure generated two time-series, one each for the aluminium profits and copper profits, consisting of 29 data points.

THE INTERVENTION AND EXPECTATIONS OF ITS EFFECT

In determining the exact time of the intervention, I believe the first rise in price in early 1974 to be the crucial one since it in effect signalled to the aluminium firms' managements that costs were rising. Jamaica unilaterally adopted a substantial tax levy in March of 1974, and most of the other bauxite nations followed its example in the next few months. Therefore, I chose the third quarter of 1974 as the intervention point.

A more difficult question is the impact of the cartel on earnings stability. Given that the earnings fluctuate in a cyclical pattern, the cartel's stabilisaton of the cost of bauxite might have an attenuating effect on earnings' fluctuations as Figure 5.1 suggests. The crucial relationship is the impact that the bauxite cartel could have on earnings from the sales of aluminium and aluminium products given that bauxite, prior to the cartel, contributed only about 8 per cent of the cost of the aluminium.

Figure 5.1 An attenuating effect on cyclical earnings

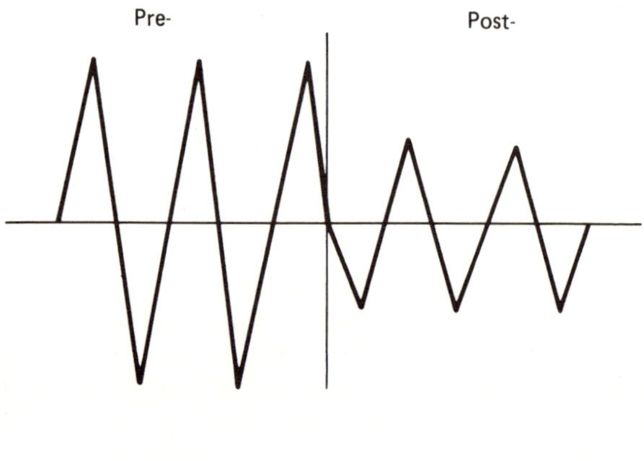

In using a quasi-experimental design any author must be particularly sensitive to the impact of other intervening variables. In particular in this case I expect inflation, which is controlled for, and the overall business cycle to have confounding effects. Since I expect the earnings of the metal producers to be affected dramatically by demand in the manufacturing sector, metal profits should fluctuate in a roughly sinusoidal pattern mirroring the general business trend. Hence, I expect to see earnings drop across the board in the 1975 recession.

If I have effectively controlled for these intervening variables, I would expect my thesis to be confirmed by the following statements:

(1) Means: $M_{A1} - M_{A2} > M_{C1} - M_{C2}$

(2) Slopes: $B_{A1} - B_{A2} > B_{C1} - B_{C2}$

where $A1$ = Aluminium profits for the 'before' period;
$A2$ = Aluminium profits for the 'after' period;
$C1$ = Copper profits for the 'before' period;
$C2$ = Copper profits for the 'after' period.

Statement (1) suggests that the intervention should result in a greater mean shift in aluminium than copper. Statement (2) suggests that there should be a greater shift in slope for the aluminium profits. The effect of both of these equations is that if the quasi-experimental design has successfully controlled all intervening variables, the intervention should produce a greater change in the experimental group in both means and slopes than in the control group. Indeed, the control group means and slopes should remain relatively constant.

Concerning the directionality of change, I expect:

(3) $M_{A2} > M_{C2}$

(4) $B_{A2} > B_{C2}$

(5) $r^2_{A2} > r^2_{C2}$,

where r^2 is the coefficient of determination, or 'variance explained'.

Statement (3) suggests that after the intervention, the average relative profits in aluminium are expected to be higher in aluminium than copper. Similarly, Statement (4) suggests that after the intervention the slope or rate of adjusted profits change should be higher (more positive) in aluminium than in copper.

Statement (5) focuses on expectations concerning earnings stability. Since the cartel stabilises the cost of bauxite, the fluctuation of relative profits around the regression line should be attenuated for aluminium as shown in Figure 5.1. With this attenuation, the variability of relative aluminium profits should be less than for copper.

FINDINGS FOR THE PRIMARY QUERY

Given the quasi-experimental design adopted here, the statistics have been calculated and listed below.[10]

$M_{A1} = 77.08$	$M_{C1} = 74.42$
$M_{A2} = 85.00$	$M_{C2} = 36.83$
$B_{A1} = 0.933$	$B_{C1} = 0.858$
$B_{A2} = -0.041$	$B_{C2} = -0.639$
$r^2_{A1} = 0.87$	$r^2_{C1} = 0.74$
$r^2_{A2} = 0.00$	$r^2_{C2} = 0.41$

Inserting these values in equations (1) and (2) gives the following:

(1)	$7.92 \not> 37.39$	Not confirmed
(2)	$0.946 \not> 1.496$	Not confirmed

Recall that in order for statements (1) and (2) to be confirmed, the left sides ($M_{A1} - M_{A2}$ and $B_{A1} - B_{A2}$ respectively) must be greater than the right side ($M_{C1} - M_{C2}$ and $B_{C1} - B_{C2}$). However, one can observe that this is not the case in each instance.

The failure of statements (1) and (2) was due to the large change in the means and betas of copper demonstrating that the design has not successfully controlled all intervening variables. Glancing at Figure 5.2, it appears that a likely intervening variable is the general business cycle. By late 1974 and early 1975 relative profits in both groups were sagging probably because of the business recession which began at that time.

Consider now equations (3) ($M_{A2} > M_{C2}$) and (4) ($B_{A2} > B_{C2}$). Substituting the appropriate values one obtains the following:

(3)	$85.00 > 36.83$	Confirmed
(4)	$-0.04 > -0.64$	Confirmed

The directionality expectations statements (3) and (4), were confirmed showing that the intervention produced a change in the direction predicted. On the average, aluminium relative profits were higher than copper. After the intervention both groups experienced falling profit rates (negative slopes) but aluminium profits were declining at a much slower rate.

Turning to equation (5) and substituting the appropriate values, one obtains:

(5)	$0.00 \not> 0.41$	Not confirmed

Contrary to the expectations of statement (5), the variability of aluminium appears to be greater at this point than that of copper. It is possible that there is a lag effect; that is, it takes a while after the intervention for profits to stabilise.

I also employed a one-year fiscal lag. The use of such a time lag seems justified by the fact that MNC's require a few quarters during which to push up the price of aluminium to their customers. It seems likely that the firms would have accomplished this gradually over about four quarters. While this process was probably gradual, I do not believe it would have been so slow as to justify multiple time lags of more than one year. In this notation M_{A2L} translates as the mean for aluminium profits after the intervention lagged four quarters. The results here were the following:

$M_{A2L} = 93.75$ $\quad$ $M_{C2L} = 19.50$

$B_{A2L} = 0.837$ $\quad$ $B_{C2L} = 0.487$

$r^2_{A2L} = 0.70$ $\quad$ $r^2_{C2L} = 0.237$

(1)	$16.67 \not> 54.92$	Not confirmed
(2)	$0.096 \not> 1.345$	Not confirmed
(3)	$93.75 > 19.50$	Confirmed
(4)	$0.837 > -0.487$	Confirmed
(5)	$0.70 > 0.24$	Confirmed

The lagged results strengthen the differences between aluminium and copper means and slopes and show that deviation from the regression line is less for the aluminium profits than for copper profits after the intervention. Again, the failure of statements (1) and (2) is due to the change in the copper groups means and betas, demonstrating that I have not taken into account the effects of intervening variables such as the business cycle. This is evident from the time-series scatter plot presented in Figure 5.2.[11]

On this graph the dark line running horizontally on the 100 per cent line represents the 1970 base line. Profits for both groups start from a low in the third quarter of 1971 due to the minor slowdown of late 1970 and early 1971 and the de-escalation of the Vietnam War (a major consumer of metal products) and rise at nearly identical rates until 1974. After that intervention year, the two lines steadily depart, both falling with the 1975 recession. But by 1976 the difference has become dramatically apparent as aluminium profits climb to new heights while copper profits continue downward.

Figure 5.2 Aluminium and copper earnings

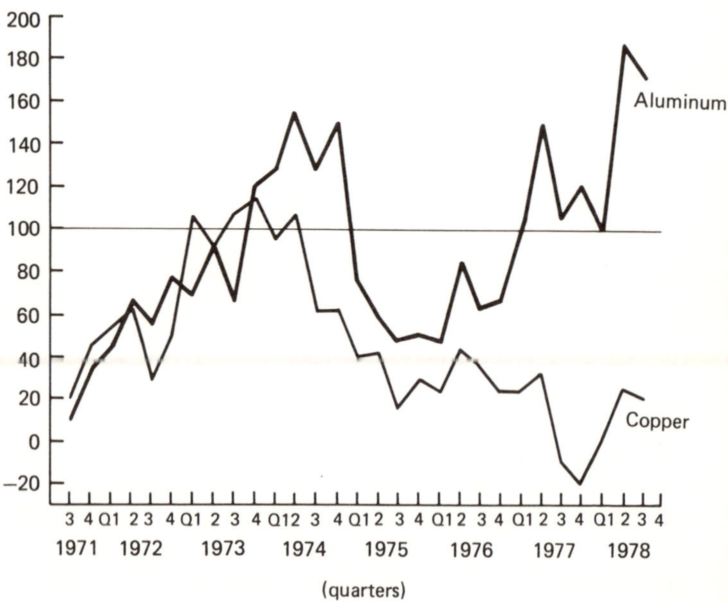

IMPLICATIONS

Given the findings since mid-1975 of an upward slope ($B = 0.837$) for aluminium profits and a downward slope ($B = -0.0487$) for copper profits, it appears that the aluminium firms have been relatively well off since the cartel was formed, at least compared with copper firms. Above all, it seems likely that the profits of the aluminium MNCs were not harmed by the imposition of the higher bauxite tax.

On the basis of the earlier expectations, this finding suggests that the aluminium firms have indeed been able to pass on price increases to the ultimate aluminium markets. This explanation seems to be further

Figure 5.3 Aluminium and copper prices

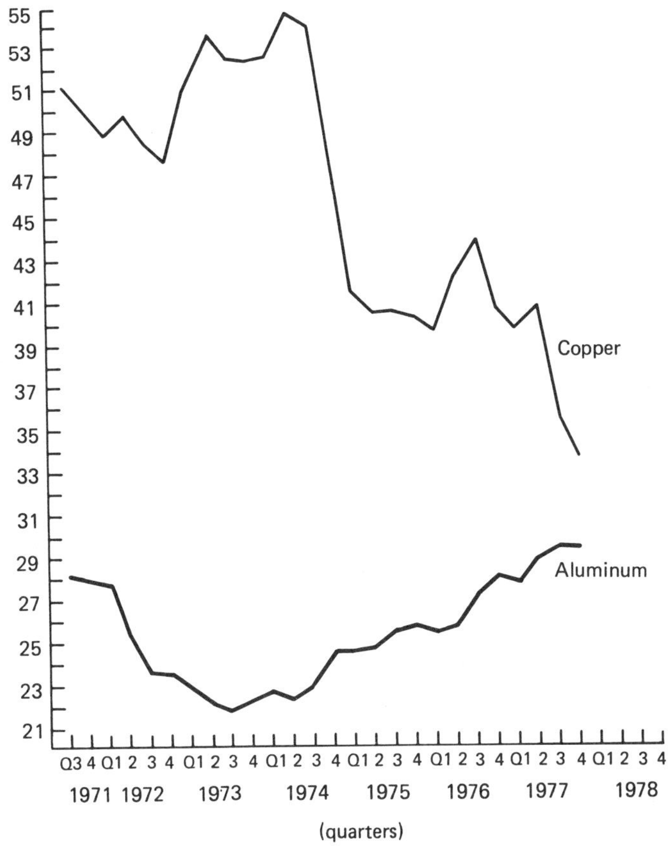

confirmed by Figure 5.3. The price chart in this figure shows the prices of a pound of copper and aluminium over roughly the same period as the last figure. These prices represent quarterly averages, which have been adjusted in terms of 1970 prices. As is clear from the graph, aluminium prices have snaked steadily upward since 1974, whereas copper prices have followed a generally downward trend.

As a result of the initiatives taken by cartel members to acquire greater control over their bauxite resources, the supply of world bauxite has seen greater co-ordination. As a result, copper prices have been more volatile and generally downward. This pattern represents but one more possible benefit to the aluminium firm of the cartel.

It should also be obvious from this state of affairs that substitution between copper and aluminium is not as great as writers have claimed (see Moran, 1974). If the two were near-substitutes, one would expect the steady increase in aluminium prices to have been offset by substitution to copper with a resultant bidding up of copper's price.

In conclusion it appears the central question of this chapter – Did the formation of a bauxite cartel benefit the MNCs? – can be answered in the case of bauxite with a qualified yes. The question remains as to the generalisability of this finding, however. No significance test has been applied to the results since the study includes nearly all the population of major aluminium and copper firms. But under what conditions might other industries benefit from governmental cartels? My best predictor of this remains my model presented on Table 1.1. This table also presents my expectations of the corporate response to the given economic situations. In this Chapter I have confirmed the first prediction: in the medium term profits did not fall. In the next chapter I examine the second prediction: the response of the companies.

6 The Response of the Multinational Companies

The success of the Caribbean bauxite producer nations on a second major NIEO front – achieving a higher rate of return on raw material exports – was traced in the previous chapter. For Jamaica and some other IBA members this involved the concerted introduction in 1974 of a tax on bauxite, in effect raising both the price of bauxite to consumers and the revenues for producer governments. As I have shown, the aluminium multinationals seem to have passed on these price increases; so much so that by 1977–8 their profits were at all time highs.

Given the evidence of higher profits in the aluminium industry, it would appear, referring to the model laid out in Table 1.1, that the situation most resembles that depicted by the bottom row of the table. According to the model, then, one would expect the companies *eventually* to adopt a passive or (as profits rise) a co-operative attitude toward the cartel. However the immediate reaction, given a one-year lag and a learning process, may be an adverse one. I attempted to document the companies' responses for the post-intervention period (1974–9) by a systematic examination of articles from business magazines (*Business Week* and *Forbes*) and business information services (*Business International, Latin America* and *Economic Intelligence Unit, Caribbean*). The findings of that examination are reported here. The expectation is to find conflictive behaviours for a year or so after the intervention turning to co-operative behaviour as higher profits are realized.

'ALUMINIUM'S BOSSES BEAMING'

As previously stated, the intervention began with unilateral action by the Jamaican government. On 18 March 1974, the government began talks with the five North American firms – ALCOA, ALCAN, Reynolds, Kaiser, and Revere – which were eventually to include a whole series of demands:

1. the renegotiation of all contracts with the firms;
2. the increasing of royalty and taxes by 600 per cent;

3. the take-over of all bauxite lands (about 200 000 acres) and control of the nation's bauxite reserves;
4. the imposition of minimum production levels for each of the firms to prevent shifting of mining away from Jamaica (eventually lowered after other bauxite nations imposed same tax); and
5. the initiation of equity participation discussions.

Faced with unilateral action, the firms adopted a combative stance:

> [They] have not taken this proposal lying down . . . the companies claim the higher tax on bauxite is unrealistic in terms of costs and investment needed to make aluminium. According to the firms, Jamaican bauxite would become uncompetitive *vis-à-vis* bauxite produced in other areas.[1]

ALCOA and Kaiser filed claims before the International Centre for the Settlement of Investment Disputes (ICSID) of the World Bank charging breach of contract. Revere (a marginal aluminium producer) refused to make its tax payments and took its case to the Jamaican courts.

Nevertheless, the events of the next few months were to vindicate Jamaica's gamble. First, by September, Guyana had substantially raised its tax levy on Reynolds' Guyana Mines Ltd, following the Jamaican example. 'The Jamaican tax formula, however, is being applied in Guyana with some adjustments for local conditions. For one thing, the Burnham Government factored in that Reynolds' shipping and mining costs are higher than in Jamaica.'[2] Second, at the same time, the new government of Surinam (the country was preparing to be granted independence in 1975) announced that it intended to follow Jamaica in demanding more for its bauxite. Third, the Congress of the Dominican Republic passed a bill to raise ALCOA's royalty payments. ALCOA mines only 5 per cent of its bauxite from the island but seemed to have responded co-operatively: 'ALCOA is showing a flexibility that should help it . . . The company may grant the Dominican Republic any advantages in price it negotiates with other countries hosting ALCOA operations.'[3] And fourth, Haiti also began negotiations for tax increases on Reynolds' local operation there, though that country provides only 10 per cent of that company's world production. Both Haiti and the Dominican Republic were invited to join the IBA along with Ghana.

Supported now by the other IBA members' actions, Jamaica realised an increase in its bauxite revenues for 1974 to $200 million (as opposed to $27 million for 1972 and 1973). It proceeded with its equity negotiations, singling out Kaiser as its first target and in late 1974 reached an agreement in principle. Kaiser seems to have been well aware of its vulnerability (it was heavily dependent on Jamaica as its bauxite source) and had long co-operated with the Jamaican government on various projects. In 1969 Kaiser established a private development corporation to channel technology and funds for development into the island economy, and proposed an innovative scheme for reversing the country's 'brain drain' problems. As a result, the 1974 agreement reflected concessions on both sides.

> the Jamaican Government will own 51% of Kaiser's existing bauxite operations. Jamaica will put up the necessary $15 million over a 10 year period at 8.5% annual interest. A new company to run this facility will be created under the direction of an executive board made up of equal representation from both Kaiser and the government. However, Kaiser will continue to manage the operations.[4]

Kaiser guaranteed Jamaica a profit return of 12 per cent. In addition, Jamaica agreed to purchase 40 000 acres of desperately needed agricultural land from Kaiser for $8 million and then lease back enough land to give Kaiser reserves of 30 years. Jamaica agreed to give Kaiser a reduction in its tax levy for three years from 8.5 per cent to 7.5 per cent. After this agreement was concluded, Kaiser decided to withdraw its case from the ICSID. 'Kaiser's President stated the new accord is "the beginning of a new and positive relationship" between his firm and the Jamaican government.'[5] This seems corroborated by the fact that shortly thereafter Jamaica invited Kaiser to participate in its talks with Mexico over a joint refiner-smelter plan.

Given this agreement with Kaiser, Jamaica turned to negotiations with Reynolds and Revere. These negotiations turned out to be quite protracted and agreement with Reynolds was not concluded till March 1977. This agreement took much the same form as the earlier agreement with Kaiser.[6]

Jamaica waited to begin the sticky issue of negotiations with the firms which had alumina refineries on the island – ALCOA, ALCAN, Alpart – until last. The negotiations with ALCOA were settled with amazing dispatch. By October 1976, in fact before the Kaiser agreement, the two had signed a pact 'which has been hailed by both parties as mutually beneficial'.[7] Basically, the plan called for the creation of a joint venture – Jamalco – to manage ALCOA's bauxite and alumina operations. The government was to hold a 6 per cent interest in Jamalco, and to have two of the seven positions on the new executive board. The provisions for the sale of ALCOA's landholdings were similar to those in the Kaiser pact.

Jamaica could take 6 per cent of the output to sale for its own alumina commitments to smelters with Mexico and Venezuela. ALCOA also agreed to withdraw its complaint from the ICSID:

> Jamalco will start off with goodwill on both sides. ALCOA's chairman billed the accord as a 'winning situation for both parties, and that's the kind of situation that works'. Manley echoed the chairman's feelings at the signing ceremony pointing out that the negotiations with ALCOA were the last to begin but the first to be completed.[8]

By contrast, protracted negotiations with ALCAN did not conclude until late 1978. Following the Jamalco formula, the government was to own 7 per cent in a bauxite-alumina joint venture creatively named Jamalcan.[9] With the conclusion of this last pact, Jamaica had successfully negotiated with each of the majors. The cases before the ICSID had all been withdrawn and equanimity appeared to reign in relations between the aluminium multinationals and the bauxite governments.

The sole exception to this melliflous ensemble was Jamaica's negotiations with the independent Revere Copper and Brass. Revere appeared to be the only firm to have stuck with an uncompromisingly querulous attitude. Revere challenged the validity of the new tax in the Jamaican Supreme Court, and refused to make its payments to the government which by January 1977 totalled US $7 million in back taxes. The Jamaican courts ruled against Revere and the firm then took its case to OPIC.

> Revere also took its complaint to the US's Overseas Private Investment Corp (OPIC) under which its Jamaican subsidiary is insured. Revere claimed that since the bauxite production levey was tantamount to expropriation, it was entitled to US $66.5 million in damages.[10]

But with all the positive euphorisms of the majors' statements, Revere's case did not come across well:

> Considering that Revere lost money in Jamaica even without the levy and that ALCOA, ALCAN and the Alpart consortium were somehow able to live with the levy, Revere's claim fell pretty flat. OPIC rejected it and the issue is now headed for arbitration.[11]

Indeed, as a marginal producer of aluminium, Revere had long had problems competing with the majors. 'Competitors say that both the Scottsboro (Alabama) smelter and the rolling mill and the Maggotty

(Jamaica) alumina refinery were all too small, too inefficient and fundamentally too badly designed to have ever paid of.'[12]

None the less, it is interesting to speculate why it was the lone minor among the aluminium firms that adopted such a conflictive position toward the cartel. A small marginal player does not have the financial and marketing resources of an ALCOA. Whereas ALCOA traditionally made its profits in fabrication and marketing, Revere was probably more dependent on finding cheap bauxite to maintain its profitability. Thus, an increase in the cost to it of bauxite took away its only advantage. Also, its more fragile finances may have made it unwilling or unable to wait and see if it could pass on the price increases. A small marginal player does not have the financial and marketing resources of an ALCOA. This again raises the question of the effect of producer nation cartels on concentration and monopoly power throughout the industry. Indeed, Bergsten suggests that this very point explains much of the co-operative behaviour of the majors: 'as in oil, the active engagement of the producing countries may create a situation which the firms were never quite able to achieve on their own – effective control over pricing and production throughout the aluminium market'.[13] Marginal producers in the industry have long had a history of trouble in competing with the majors. For Revere, the cartel appears to have been the straw that broke the camel's back. In the second quarter of 1980, Revere announced it was seeking a buyer for its Jamaica operations.[14]

Indeed, since the formation of the IBA, the majors have changed their policy on pricing. '"We have left the path of (market) pentration pricing for a path that emphasises financial performance," explained John L. Diedrich, ALCOA's general sales manager'.[15] Consequently, ALCOA boosted its price several times in 1977 and 1978; and 'what ALCOA does, the other aluminium producers will follow (and did). In such a position of strength, ALCOA is telling customers, in so many words, if they want aluminium . . . they will have to pay for it'.[16]

By early 1979 the result was clear. With profits climbing ever higher, ALCOA chairman, W. George seemed to be near attaining his predicted (back in February 1977) 14 per cent return on investment, a figure unheard of for ALCOA since its old monopoly days. '. . . domestic producers have chosen to hold back expansion, run at near-capacity, and jack up prices as far as demand – and the law – will allow'.[17] However, the majors have few plans for expansion and with aluminium supplies growing short, one wonders why no new entrants are being attracted to the field. *Business Week* lays the blame on higher costs and bauxite. '*An assured supply of bauxite has been a key factor ever since the formation of the cartel-like International Bauxite Association*'[18] (my emphasis). With this greater profitability and predictability *Forbes* has been running headlines to articles such as 'Aluminium's bosses are Beaming', and 'Aluminium Turns to Gold'.

To summarise, this chronicle has traced the reaction of the firms to the cartel formation in Jamaica. The initial reactions of all the companies were intransigent and conflictive. In particular, the companies responded by beginning legal proceedings against government and making bellicose public statements. In addition, Revere decided to withhold taxes. Shortly thereafter, all four of the majors – ALCOA, ALCAN, Kaiser, and Reynolds – dropped their conflictive responses in favour of co-operative actions. These positive reactions included offering concessions in their bargaining positions, by allowing partial government ownership of the mines, and issuing conciliatory statements. From this narrative I conclude that the ultimate response of the major firms was co-operative as predicted on the basis of the last row of Table 1.1.

IMPLICATIONS: THE 'PROFIT-SHARING CARTEL'

Some analysts have compared the bauxite cartel to OPEC. In both cases, the producer governments had to successfully challenge the control of the majors. In the case of bauxite, Guyana demonstrated to the majors that their positions were not invincible, playing much the same role as Libya did in relation to the oil industry:

> The unilateral seizure by Guyana of its subsidiary in 1971 had something of the catalytic impact in disrupting the traditional dominance of the companies that Libya's oil nationalization and price demands did during the same period.[19]

As with oil, the bauxite price increase was accomplished by individual governments following the example of a price leader:

> The scenario as it developed represents 'price leadership' by Jamaica, followed individually by other countries, rather than pure cartel action. This is exactly what happened in oil in 1971, when Libya and the Persian Gulf states began the oil offensive by 'leapfrogging' each other's tax demands.[20]

Finally, in both cases, governments identified and pressured the weak-link in the companies' chain. Just as Libya began its renegotiations with the minor oil companies, so Jamaica began its renegotiations with

Kaiser, the one company most dependent on its bauxite. Only after Kaiser had begun to cave in to its pressure did Jamaica tackle the top MNC, ALCOA.

But, these comparisons apart, the bauxite agreements are not entirely the same as for the OPEC cartel. In negotiations over equity, Jamaica pursued participation rather than expropriation. Yet the real breakthrough for the bauxite countries was in indexing the price of bauxite to the price of aluminium. Jamaica originally demanded a tax of 8 to 8.5 per cent of the going price for an aluminium ingot on each ton of bauxite mined by the firms, but lowered the rate to 7.5 per cent after successful conclusion of the equity agreements. Perhaps the original motive for linking the tax on bauxite to the final price of aluminium was to end the past practice of relying solely on the companies' estimates of the worth of the bauxite for tax purposes (given the absence of a world bauxite market).

But the linkage had another important aspect: as with most indexing schemes, it has a built-in escalator. Given the nature of the aluminium industry, higher prices mean higher profits for the majors and higher tax revenues for the cartel. In effect, the two groups appear to be co-operating to exercise monopoly power. The result resembles a 'profit-sharing' scheme rather than confrontation between the companies and the governments.

> To promote such a result (greater co-ordination with the firms), IBA explicity seeks 'partnership' with the companies. Indeed, Jamaican officials advocate higher profits for the firms in the near future. The producing countries insist on being the senior partners, but clearly recognize that it behooves them to use the companies rather than treat them as enemies.[21]

As the price of aluminium has risen through the 1970s. Jamaica's 'cut of the take' has increased steadily. But for the firms, gross profits are still profits, particularly when the cartel is in part responsible for those higher profits.

> Much of the initial adverse reaction is attributable to the speed with which Jamaica moved, and the need of the companies to avoid repeating the mistake of the oil companies which appeared to cave in too quickly in 'abandoning the consumer interests'. IBA staff and company personnel attend each other's parties and seem on wholly friendly terms.[22]

What I am calling a 'Profit-Sharing Cartel' in this case is very similar to the role which Vernon (1977) sees the MNCs playing as 'tax collector' for the developing nations. By this phrase he refers to the attempts by some LDCs to use MNCs to raise the prices of their commodities, with a bigger share of the increased return going to both parties. In Vernon's terminology, the aluminium firms are definitely acting as the 'tax collectors'. As was shown previously, the companies appear to have little difficulty in passing on the price increases from bauxite to aluminium. With this rise in price, both parties have reaped bigger takes, as the companies' higher profits and governments' increased revenues would seem to suggest.

If the aluminium firms serve as tax collectors for the bauxite nations, then it is consumers and other businesses in the US and other industrialised countries which must pay the new rate of 'tax'. If the consuming nations continue to respond in the disorganised, passive manner that followed the initial rounds of the OPEC oil increases, then the situation is not likely to change in the near future.

Compared to oil, the revenues being redistributed by the bauxite levy are not of sufficient magnitude to attract the attention of the consuming nations. Indeed, the amounts of revenue are significant only for the governments of the various IBA members. For example, during the 'heyday' of the bauxite cartel – roughly 1974–9 – the Jamaican bauxite levy added an average of US$200 million annually to that government's coffers. This figure represents a quarter to a third of the Jamaican government's annual revenue. It is not entirely clear how this redistributed wealth benefited Jamaica – other than in an expansion of government services and civil service positions. Undoubtedly, it paid for the partial nationalisation of the companies. But it did not produce the sort of 'economic miracle' that oil produced for Mexico or other OPEC members. In part this absence of a windfall in the Jamaican economy was a function of its smaller size, other chronic economic problems, and even a hurricane which devastated the 1979 sugar crop.

Another contrast with oil was apparent in 1979. While OPEC members again increased the price of oil substantially that year, the Manley government actually reduced the bauxite levy from 7.5 per cent to 6.0 per cent of the price of a finished aluminium ingot. Given the disastrous economic conditions of 1979, the Jamaican government hoped to give the companies an incentive to expand their bauxite production on the island. This hope was ill-founded for a number of reasons. First, an expansion of production violates the logic of cartel which calls for production cutbacks to maintain price. Second, with the new increases in energy costs the companies were now much more sensitive to expanding aluminium production only in countries with lower energy costs such as Brazil and Venezuela. And finally, even with this cutback in cost, Jamaican bauxite was still more expensive than bauxite from IBA free-riders, Australia and Guinea.

With Seaga's election in 1980, many commentators predicted the demise of 'Manley's' bauxite levy. Yet despite Seaga's preference for 'free market' ideas, the Jamaican levy has persisted under his government. As of early 1986, it has survived re-evaluations and renewals at the 6 per cent of an aluminium ingot price rate. Chapter 7 will document recent developments in the bauxite/aluminium industry.

CONCLUSION

This chapter has found evidence in the bauxite cartel of the 1970s to support the simple model of company response given in Table 1.1. This model cannot alone predict the ultimate success or failure of the cartel, however. Clearly other factors such as member solidarity (or its lack thereof in the cases of Australia and Guinea) are important in determining the longevity of the levy. But the company reponse is none the less a crucial contributing factor by itself.

My model is simply an elaboration of the fact that corporate behaviour is determined by the quest for profit and stability. In the second half of the decade of the 1970s, conditions within the bauxite/aluminium industry existed to allow the operation of such a 'profit-sharing' cartel. This type of cartel deserves more attention for it represents a unique vehicle for the redistribution of global wealth using the MNCs as conduits (or 'tax collectors') for the Third World. I suspect that application of this analysis to the oil industry would reveal an even clearer example of 'profit-sharing'.

In the final chapter, I will discuss the application of my model to two other commodities – uranium and bananas – and show its limitations in the case of bauxite. As with OPEC, by the 1980s the conditions which had fostered success in the late 1970s were rapidly eroding.

7 Limits and Scope and Conclusions

This study has shown the relevance of the aluminium MNCs to the formation of the bauxite cartel of the mid-to-late 1970s. I have suggested that the co-operation of the international companies is crucial in any producer raw material cartel. I presented a model for predicting the response of the MNCs (co-operative, passive, or conflictive) based on the underlying characteristics of the particular industry in Chapter 1. In situations of demand inelasticity and oligopoly, the firms may be most open to government cartel efforts.

Having achieved some success with this mode of analysis in the case of aluminium, I ask whether it is applicable to other commodities. No sooner had OPEC achieved its spectacular price-hikes of 1973–4, than it was being heralded as the 'exceptional' commodity for such cartelisation. But, as I have demonstrated, in 1974 the Caribbean bauxite producers achieved their own OPEC-like price increase. I believe my 'profit-sharing' cartel analysis would also be appropriate for the case of oil (demand inelasticity, oligopolistic structure).

In this chapter I will examine two other commodities where cartels have been attempted, one 'exceptional' – uranium – and one failure – bananas. In both cases, the role of the MNCs was crucial. My somewhat speculative analysis here will be narrative and preliminary rather than extensive and conclusive. I will then conclude with a discussion of the bauxite cartel in the 1980s.

THE URANIUM CARTEL – ANOTHER EXCEPTION

Faced in 1971 with both chronic oversupply and soft prices, the governments of two major uranium suppliers, Canada and France, began actively to promote a uranium cartel complete with production quotas and assigned prices. From the beginning the uranium companies were assumed to have a major role. Again we can apply the Chapter 1 model to predict company response.

Demand for uranium would appear to be quite inelastic, even more so than its energy alternatives. Nuclear reactors involve major capital investments. Given the problems of contamination, reactors are not easily closed or 'moth-balled'. Once committed to nuclear power a utility company appears locked in as a uranium price-taker.

Furthermore, many governments have been committed to a nuclear energy programme as a good in itself. Even before the oil embargo, governments promoted nuclear reactors for energy autonomy and 'hi-tech' reasons. Government sponsorship with its bureaucratic inertia builds in demand inelasticity.

The uranium industry's structure is unique on a number of counts. First, it is obviously a very new industry. Most of its development has come since the Second World War. Hence, there is little or no history of MNC diplomacy specific to uranium. Second, given the security implications of nuclear power, national governments have been much more involved from the start than in either oil or aluminium. In Canada, both Uranium Canada Ltd and Eldorado Nuclear Ltd have been crown corporations. Furthermore, for most of the industry's history the US has been a closed market to uranium suppliers jealously guarded by the American Atomic Energy Commission (AEC). None the less, industrial concentration is fairly high. In 1972, of non-US uranium sources, Rio Tinto-Zinc Corp. Ltd (London) controlled 17 per cent, Denison Mines (Canada) 13 per cent, Anglo-American (S. Africa) 5 per cent, Imetal and other French companies 15 per cent. Among the Rothschild's Imetal, Oppenheimer's Anglo-American, and Rio Tinto-Zinc there were numerous other interlinkages in ownership, agreements, and directorates. Thus, I conclude competition is not high in among the uranium MNCs.

Under the model, I would therefore expect that when meetings to form the cartel were held in 1972, the companies would on the whole respond positively. Indeed, the Paris meetings of 1–4 February 1972 were a two track affair with Canadian government officials calling together officials from Australia, France, and South Africa while corporate representatives of the major firms met, hosted by the French companies. The meetings in effect were co-ordinated by the government-owned status of some of the firms (e.g., Canada's Eldorado). By 1 June 1972, the club members had agreed to both production shares and a formula of price increases, more than either the oil or bauxite cartel accomplished. In the five years that the first cartel was to run prices soared from $5.30 per lb in 1972 to $26.00 per lb in 1975. In fact by 1975 the price was so far above the agreement, the cartel had become superfluous. Companies and governments alike benefited from this five-fold price increase and guaranteed production.

The one problem company from the Canadian government's perspective was the American Gulf Oil affiliate. Arguing that its participation in such an elaborate cartel would lead it foul of US anti-trust authorities, Gulf was reluctant to participate until the Canadian government compelled it to join (or lose its Canadian production). In this case, Gulf proved to be prescient. In 1976 Westinghouse (US), a major consumer of uranium, filed a suit against the club, charging price fixing, and beginning the process of public exposure of the cartel.

Despite some similarities, we can see in the uranium example features which distinguish it from the bauxite cartel. While a producer governmental cartel, the governments involved here were all of relatively developed countries. One can therefore expect considerably more regulatory clout over the companies. In part, this advantage was offset by stricter anti-trust laws which for example, forced the Canadian government into some fairly acrobatic legal manoeuvreing with itself over its own Anti-Combines Act. Furthermore, as developed countries the international income redistribution aspects are much less salient. The 'tax' collected here is basically levied on consumers in their own domestic markets. Indeed the impetus for the cartel was not the NIEO but the natural extension of a domestic regulatory function to the international level. The Canadian and French governments responded to price instability and oversupply among their uranium producers as they might have responded to the overproduction of milk domestically. Hence in some aspects this type of governmental cartel might better be called a 'Regulatory Cartel'.

None the less, as with bauxite, we see here again the importance which governments placed on having the co-operation of the companies. As with bauxite, a few companies demurred on anti-trust grounds but 'gave-in' when pressured. Perhaps the show of resistance was more to establish an alibi in case of future litigation.

WHERE THE COMPANIES FIGHT BACK – THE BANANA FIASCO

Most agricultural commodities would not fall under the conditions described by the bottom row of my model. Even in the case of bananas where the industry is highly concentrated around three majors – United Brands, Del Monte, and Castle and Cook – the requirement of demand inelasticity is not met. Bananas are not a requirement of most people's diets and if the price goes up significantly, they will simply stop buying them. In such an industry the model would predict that the MNCs would not be able to pass on price increases to consumers and thus profits would suffer. It should be no surprise then to find that the banana multinationals have sabotaged every attempt by the banana producers to form a cartel.

At times this banana battle has achieved the heights of soap opera. On 3 February 1975 President Eli Black of United Brands committed suicide by leaping from the 44th floor of the Pan Am building in New York. Shortly thereafter a Securities Exchange Commission investigation discovered that United Brands paid out $1.25 million in bribes in Honduras. In 1974, Honduras and other banana producers formed a banana association the UPEB (Union de Prises Exportadores de Banano) and introduced a $1 per box tax on bananas. The companies successfully fought back with a divide and rule strategy. With the assistance of the bribe, United Brands got Honduras to lower its tax to 25¢. When Panama refused to drop the tax, United Brands closed its operations there. Ecuador and Colombia refused to join in the cartel. Certainly the UPEB example shows how unmitigated hostility from the MNCs can hamper any cartel attempt.

Certainly for a product with an elasticity of demand as great as bananas, cartelisation is questionable under any conditions. Even products where the demand appears inelastic may none the less prove to be more elastic at very high prices. For example, high oil prices made 'expensive' alternatives and marginal oil sources viable over time. As the aluminium MNCs pushed up the price of aluminium in the late 1970s, they may well have discovered the same phenomenon.

ALUMINIUM IN THE 1980s

The predictions of my model can also be affected by long-term shifts in the 'givens' of an industry – demand elasticity and industry structure. Recent evidence suggests that aluminium is undergoing a profound transformation which challenges the assumption of demand inelasticity and non-competition. Though the Jamaican bauxite levy persists, the long-term viability of the 'profit-sharing' cartel has been thrown into question by a number of developments in the 1980s.

First, profit-sharing is premised on there being profits to share. Following the trend in industrial activity, the aluminium industry entered a depression in 1981 from which it is only now beginning to escape. This collapse in demand for aluminium was for the most part due to the general downturn in industrial activity of the early 1980s. Also there is some evidence that the aggressive pricing policy has begun to result in substitution in the auto and container industries towards the new 'super plastics'.[1] In the long term, at high enough prices, demand for aluminium may yet prove to be more elastic than expected. Profitability has been hurt by the further leap in the electric energy costs following the 1979 oil price hike. Increased energy costs have contributed much more to the cost of an aluminium ingot than the bauxite levy. As a result of all these factors by the second quarter of 1983, most of the majors had reported three consecutive quarters of losses. In 1983 ALCAN's president announced that his firm 'had lost money last year for the first time in fifty years'.[2]

Loss of profitability tends to increase the scramble for markets and competition among the majors. Indeed, by 1983 the majors' solidarity on pricing had ended. ALCAN adopted a 'free rider' position selling its aluminium just under the world price. ALCAN's aggressive sales strategy 'prompted some resentment from its three competitors in the Big Four North American aluminium producers, who charge that ALCAN's prodigious production will delay the market's upturn'.[3] ALCAN decided to try to take advantage of its position as low-cost producer to weather the recession. As a result, by 1983 it operated at 88 per cent capacity while ALCOA had to cut back to 70 per cent and Reynolds and Kaiser dropped to an astounding 23 per cent and 19 per cent respectively. Kaiser has dropped so low that by 1986 it was desperately fending off a hostile take-over bid by the Frates Group.

Thus, by the end of the 1970s, there appeared to be a profound shift in the structure of the industry. As one *Wall Street Journal* headline put it in 1982 'ALCOA's Dominance in Aluminium Industry Wanes as Rivals Grow, Markets Get Tight'[4] Aluminium smelter capacity grew rapidly in the 1970s producing oversupply in the 1980s. The Majors' position was weakened by this expansion as their control of non-communist smelter capacity dropped from 80 per cent in 1970 to 40 per cent by 1982.[5]

Perhaps the clearest sign that the majors have lost control of the industry has been their inability to prevent aluminium's listing on the major commodity exchange. In 1978, aluminium was listed with the London Metal Exchange and then, December 1983, over the companies' protest, it was listed on the Commodity Exchange of New York (COMEX). Traditionally the price of an aluminium ingot was determined by the aluminium majors: now it will be determined by the commodity markets. The majors fought and lost this battle. As one ALCAN executive put it, 'I've always thought that one of the greatest virtues of aluminium was the price stability it enjoyed. Now that it's on

COMEX, you can imagine the loss of stability we're all in for.'[6] In effect, aluminium prices may now be expected to fluctuate like any other commodity.

In the production of bauxite, the Caribbean members of IBA have found their positions eroded by four 'free riders' to the cartel. Guinea and Australia are both IBA members but Australia has no bauxite levy and Guinea has set its levy well below the Caribbean rate. Australia's lack of levy can be traced to conservative Fraser administration's ideological commitment to 'markets'. Guinea's decision to undercut the cartel seems based on classical free rider behaviour – to price just under the cartel price and thus seize a greater market share. Venezuela and Brazil have both just integrated backwards from their existing smelters into bauxite mining and neither are members of IBA. This move is rational for both of them – it gives them a fully integrated industry and serves the balance of trade of each by cutting an important import – but it further weakens the cartel, particularly if Brazil and Venezuela export their surplus. The IBA has actively sought Brazilian membership, but for the moment the Brazilian Government wants only observer status.

Indeed, Brazil is widely touted to be the aluminium superstar of the future. With lots of bauxite, abundant cheap hydroelectric power, and a large domestic market, Brazilian smelting capacity has jumped from 299 000 tonnes in 1982 to 550 000 tonnes in 1985 and is expected to double again by 1990 to 1.2 million tonnes. If Brazil achieves that figure, it would become the market-economies' second or third largest producer. Of course, the aluminium MNCs have been scrambling over each other to get into the Brazilian industry (a mixture of government and private companies).

Over time, some of the aluminium majors have located their new production in free rider territory and closed some production in the Caribbean. They have expressed dissatisfaction with the bauxite levy. Clearly the conditions which promoted co-operation in 1976 do not remain in 1986. As my model predicts: as the industry structure has changed away from conditions favourable to cartelisation, the firms have become more hostile to the cartel.

In August 1985, shortly after Reynolds' decision to leave Jamaica, Wesley Hughes of the Jamaican Bauxite Institute summarised the reasons for the problems of both the Caribbean and the IBA:

1. a sharp downturn in consumption;
2. a shift in aluminium smelting away from the US Gulf Coast, Jamaica's traditional buyer;

3. the loss in the world market share controlled by the aluminium MNCs;
4. more worldwide government ownership; and
5. 'the demise of multinational control over the price of metal'.[7]

His last three points seem to differ sharply from the traditional NIEO view of the MNCs as clear enemies of the Third World.

The present weakness of the IBA suggests an important point in the use of my model. MNC acquiescence is a necessary *but not sufficient* condition for cartel success. Political will and solidarity on the part of the cartel member governments remain the ultimate factors. It is impossible to keep the MNCs and the cartel in line if some members decide to cheat (Guinea and Australia).

THE TAKE-THE-MONEY-AND-RUN CARTEL

The seeming collapse of OPEC in early 1986 casts a shadow over the prospects for future cartels. Commodity prices across the board are now at an all time low. Yet the oil example poses an interesting question – what must a cartel accomplish to be considered successful? In 1973 and again in 1979 the oil producers generated a massive redistribution of revenues in their favour. Should Saudi Arabia deny its success in capturing revenues far above what the market would have given in that period?

This pattern of monopolistic behaviour could be referred to as a 'take-the-money-and-run cartel' of which J.P. Morgan's US Steel monopoly provides an example. When financier Morgan put together his steel monopoly at the turn of the century, he had no intention of maintaining an ALCOA style fifty year monopoly. He wanted his profits immediately. US Steel's pricing and profit policies eventually attracted rivals into the steel industry, but not before US Steel had extracted many years of monopoly profits. In effect, US Steel traded market share for high profits. Eventually, US Steel dropped to competitive prices to halt the erosion of its market share. Only at that point did it begin earning the lower profits it would have always earned, had it priced competitively from the beginning. For Morgan, the monopoly was successful just for producing above normal profits for a limited time.

There is evidence that ALCOA knows it is now pursuing such a US-Steel-type strategy. As late as 1982, ALCOA chairman W.H. Krome George continued to defend the aggressive pricing policy he had introduced in 1975. He still maintained that ALCOA would, should be 'the most profitable company, not the largest'.[8]

Now some will argue that while the oil and bauxite cartels gained short-run profits they lost market share. The revenue of big profits over smaller production may not have equalled the lost small profits on a larger production. This is basically an economic question and the evidence for Jamaica does not seem to bear it out. Government revenues were much higher in the post-1974 period. Furthermore, the argument that Jamaican production would have expended greatly if the levy had not occurred is doubtful at best. The factors that make Australia and Brazil attractive would have drawn investment to them with or without the IBA. Jamaican production might well be near its present production level even without the levy. The tax, while very significant for Jamaica, is a very small cost factor to the companies, compared to energy costs, for example. And since 1982 the levy can be claimed as a foreign tax credit in the US!

There is one final and compelling argument for mineral and oil producers to go for profit over market share. Oil, bauxite and uranium are non-renewable resources, unlike bananas. Production foregone today is not production foregone forever. Eventually the reserves of the mineral – at least the easily expolitable reserves – will run out. Does it not behoove the government therefore to gain the highest price while the good lasts?

Certainly the bauxite levy has not yet generated an economic miracle for Jamaica, but how much worse off would the island be if it had not had the additional revenue? How much worse would its balance of payments problems have been if it had not had these large company payments to offset its escalating oil bill? By pricing its bauxite so high Jamaica has lost some short-term production: it has dropped from a record high of 15.2 million tonnes in 1974 to 8.6 million tonnes in 1984. But given the oil price increase and recessions of this period, Jamaica may well have lost this production anyway.

In 1983 the Jamaican government itself evaluated the benefits and costs of the levy on the eve of its decision to renew it. Overall, the evaluation was positive despite the loss of market share. In 1982 the levy contributed directly US$135 million of the government's revenues. Under the pre-levy method of calculation (45 per cent tax on profits), this would have been a mere $15–20 million. Further, the government would have been dependent on the companies' reports of profits and concerned about possible transfer pricing. The information and experience gained by the IBA must also be added as a benefit.

Everything considered, the Seaga regime decided in May 1984 to continue the levy at the going rate of 6 per cent of the price of aluminium.

Commodity markets appear at present to be in universal disarray. Yet commodity production tends inevitably to be cyclical. Depressed markets shake out marginal producers and increase the market power of the survivors. OPEC and other cartels will certainly rise again. We have not seen the last of cartelisation attempts nor this specific form; the profit-sharing cartel.

Notes and References

1 Introduction

1. International Economic Studies Institute, *Raw Materials and Foreign Policy*, 1976, serves as an example of this line of analysis.
2. An example of this line is Mikdashi (1976).
3. Pindyck (1977), p. 356.
4. Wyant (1977), p. 2.
5. Ibid., p. 2.
6. Caves (1982), p. 234.
7. See for example Samuelson (1973), p. 483.
8. 'The Charter of Economic Rights and Duties of States', *UN Monthly Chronicle*, 12, no. 1 (June 1975), p. 108–18.
9. Mikdashi (1976), p. 116.
10. Particularly if co-operation creates a stronger monopoly over source than had existed before.

2 The Aluminium Industry: A Descriptive Profile

1. Sampson (1975), Turner (1978) and Wyant (1977) to name but a few.
2. Information in this section is based on *Aluminium by ALCOA* (1969), Brubaker (1967), Farbin and Reibsamen (1969), Narvin (1978).
3. Information in this section is based on *Aluminium by ALCOA* (1969), Brubaker (1967), Farbin and Reibsamen (1969), Carr (1952), Peck (1958), Stocking and Watkins (1946), and Wilkins (1974).

3 A History of the Aluminium Industry's Cartel

1. The terms 'foregin policy' and 'international diplomacy' are applied to MNCs in the manner adopted by Alger (1977) in speaking of non-governmental actors.
2. The company history written by ALCOA public relations man, Carr, is particularly amusing on this point:

   > ALCOA's management decided that it was not equipped to develop foreign markets, and that this could be done adequately only by a foreign corporation devoting its major activities to the task . . . This voluntary decision on the part of ALCOA to devote itself to the United States market was, as we have already said, inspired to some extent by the amazing acceptance aluminium received in this country in the twenties (Carr, 1952, p. 191).

   No mention is made of the cartel or the Consent Decree. It strains credibility that ALCOA spun off a good share of its business for no other reward than to create another competitor.

3. Hoffman (1974), p. 147.
4. In his conviction of ALCOA as a monopolist, Judge Hand ruled that 90 per cent of the ingot market represented a monopoly. Dealing then with the slippery notion of what monopoly power is, the Judge suggested, 'It is doubtful whether sixty percent (of the market) would be enough'. (Quoted in Peck (1958), p. 12.) Hence the reification of 60 per cent.
5. The company's account of these events again attributes ALCOA with the loftiest of intentions:

   > . . . the Company's fight against subsidies and other socialistic suggestions, plus its liberal attitude toward competitors who acquired the plants, did much to bring about the present competitive conditions in the industry (Carr, 1952, p. 264).

6. Wilkins (1974), p. 297
7. Marvel (1980).
8. *The Economist,* 1978, p. 97.
9. Ibid.
10. In brief, Lenin argued the capitalist firms like capitalist nations are too competitive to forge long-term agreements. Kautsky in describing 'ultra-imperialism' emphasised co-operation among firms in global cartels. For more see Holloway (1983).

4 The Third World Bauxite Producers

1. Wilkins, p. 11.
2. US Bureau of Mines, August 1954.
3. Wilkins, p. 254.
4. Graham, p. 96.
5. Graham, p. 98.
6. 'Guyana to Nationalize ALCAN Subdiary, after Failure to Work out Joint Venture', *Business International Latin America (BILA)*, 1971, p. 74.
7. Graham, p. 101.
8. Graham, p. 101.
9. 'Jamaican Guyanese Politics on Bauxite – Contrasting Expression of Nationalism', *BILA*, 1971, p. 239
10. *Guyana: A Decade of Progress*, 1974.
11. *Metal Statistics*, 1976.
12. 'Nationalism Closeup: How Guyanese Takeover is Faring Five Years Later', *BILA*, 1976, p. 199.
13. Graham, p. 204.
14. Graham, p. 103.
15. Graham, p. 105.
16. 'Towards Increased Co-operation in Bauxite Development', *Proceedings of the International Conference on Bauxite*, Jamaica, December 1980, p. 123.
17. Further information see Mitchell (1968), Kuper (1976), Girvan (1971).

18. Girvan (1971), p. 15.
19. Kuper (1976), p. 17.
20. Mitchell (1968), p. 135.
21. Kuper (1976), p. 129.

5 The Impact of the Bauxite Levy on Company Profits

1. *General Assembly Official Record*, 27th session.
2. Clarfield *et al.* (1975), p. 123.
3. Ibid., p. 124.
4. For complete profiles on each metal see Steven Holloway, *Do Multinational Corporations Benefit from Government Cartel Formation?* unpublished Doctoral Dissertation, Ohio State University 1979, ch. 3.
5. Intersubstitution of the two metals was considered, but rejected as not important to the study for several reasons. First, in many of aluminium's container, aerospace, and nautical uses, copper's chemical properties make it unfit as a substitute. Even in the area of electrical uses, the two metal's properties are significantly different (conductivity and weight) and prevent them from being considered perfect substitutes. Second, whatever substitution might occur, it is unlikely to be observed in the short time frame of the study, 1974–8. And third, to the extent that consumers might switch to copper after aluminium prices increase, profits in aluminium would decline, and the hypothesis would be that much easier to reject. As will be shown, the study found no evidence of intersubstitution.
6. The Copper Industry has been extensively treated elsewhere; see Moran (1974) and Navin (1978).
7. At this point some reference should be made to the variable 'risks'. It is customary in the business literature to discuss 'profits' in the context of 'risks'. The normal state of affairs is that the higher the risk, the higher the profits must be before an investment will be made. However, the metal industry is a mature business in which the size of the major firms and the lengthy past experiences of those firms has diminished the importance of risk, as witnessed in the traditionally low but stable profits made in the industry. Furthermore, 'risk' has been excluded as a variable due to the difficulties in finding a reliable, quantitative measure of it.
8. Narvin (1978), p. 271.
9. See Holloway (1979) for more on the calculation of the index.
10. The statistics were generated from an SPSS programme using the regression calculation routine.
11. I was unable to find a control for the business cycle. Several economists consulted agreed that no reliable measure of the business cycle exists.

6 The Response of the Multinational Companies

1. 'Jamaican Move Signals New Era for Bauxite Firms', *BILA*, 1974, p. 173.
2. 'Guyana's Tax Levy on Reynolds' Operation Foreshadows Worldwide Trend', *BILA*, 1974, p. 134.

3. 'Other Caribbean Bauxite Countries to Raise Ante', *BILA*, 1974, p. 314.
4. 'Jamaican Move on Kaiser Equity Unveils Government's Aim', *BILA*, 1976, p. 119.
5. 'Jamaica Reaches Major Accords with Two Bauxite Producers', *BILA*, 1976, p. 119.
6. 'Jamaican Accord with Reynolds Metals Co. is One More Positive Step', *BILA*, 1977, p. 116.
7. 'Jamaica's Accord with ALCOA Marks Major Breakthrough . . . ', *BILA*, 1976, p. 344.
8. Ibid., p. 344.
9. 'ALCAN Deal in Jamaica . . . ', *BILA*, 1978, p. 329.
10. 'Jamaican Accord with Reynolds Metals Co. . . . ', *BILA*, p. 117.
11. 'Staying on Top', *Forbes*, July 1, 1977, p. 56.
12. Ibid., p. 54.
13. Bergsten (1976), p. 16.
14. 'Revere to Sellout?', *Economist Intelligence Unit*.

7 Limits and Scope and Conclusions

1. Agis Salpukas, 'A New Hunt for Markets at Reynolds', *The New York Times*, 22 September 1981, D6.
2. 'How ALCAN endured the recession', *Maclean's*, 4 April 1983.
3. Ibid.
4. Amal Nag, 'ALCOA's Dominance in Aluminium Industry Wanes . . . ', *Wall Street Journal*.
5. Ibid.
6. 'Aluminium producers unhappy with listing', *Chronicle Herald* (Halifax), 15 December 1983.
7. 'Up Against the Wall', *Latin America Commodities Report*, 30 August, 1985, p. 2.
8. Nag, 'ALCOA's Dominance'.

Bibliography

ALGER, CHADWICK (1977) 'Foreign Policies of US Publics', *International Studies Quarterly*, vol. 21, no. 2: 277–318.

ALUMINIUM ASSOCIATION (1974) *Aluminium Statistical Review.*

ALUMINIUM COMPANY OF AMERICA (1969) *Aluminium by ALCOA.*

AMERICAN BUREAU OF METAL STATISTICS (1978) *Non-Ferrous Metal Data 1977*, New York.

BANKS, FERDINAND E. (1974) *Bauxite and Aluminium: An Introduction to the Economics of Nonfuel Minerals* (Lexington: Lexington Books).

BERGSTEIN, C. FRED (1976) 'A New OPEC in Bauxite', *Challenge*, July – August: 12–20.

BERK, RHEA *et al.* (1982) *Aluminium: Profile of the Industry* (New York: McGraw-Hill).

BIERSTECKER, THOMAS (1978) *Distortion or Development* (Cambridge: MIT Press).

BRUBAKER, STERLING (1967) *Trends in the World Aluminium Industry* (Baltimore: Johns Hopkins Press).

CAMPBELL, D. and J. STANLEY (1966) *Experimental and Quasi-Experimental Designs for Research* (Chicago: Rand McNally).

CAMPBELL, DONALD (1969 'Reforms as Experiments' in *American Psychologist*, vol. 24, no. 4: 409–29.

CAMPBELL, DONALD and THOMAS COOK (1976) 'The Design and Conduct of Quasi-Experiments and True Experiments in Field Setting', in M.D. Dunnette (ed.) *Handbook of Industrial and Organizational Research* (Chicago: Rand McNally).

CAPORASO, JAMES and LESLIE ROOS (1973) *Quasi-Experimental Approaches to Social Sciences: Perspectives and Problems* (Evanston: Northwestern University Press).

CARR, CHARLES (1952) *ALCOA: An American Enterprise* (New York: Holt, Rinehart & Winston).

CAVES, RICHARD (1982) 'International Cartels and Monopolies in International Trade' in John Letiche (ed.), *International Economic Policies and their Theoretical Foundations* (New York: Academic Press).

CHEN-YOUNG, PAUL (1973) *Report on Private Investment in the Caribbean* (Kingston, Jamaica: Atlas Publishing).

CLARFIELD, KENNETH *et al.* (1975) *Eight Mineral Cartels* (New York: McGraw-Hill).

FARBIN, PHILIP and G.G. REIBSAMEN (1969) *Aluminium: Profile of an Industry* (New York: McGraw-Hill).

GALBRAITH, JOHN K. (1952) *American Capitalism* (Boston: Houghton Mifflin Company).

GALBRAITH, JOHN K. and NICOLE SALINGER (1978) *Almost Everyone's Guide to Economics* (Boston: Bantam).

GILPIN, ROBERT (1975) *US Power and the Multinational Corporation* (New York: Basic Books).

GIRVAN, NORMAN (1971) *Foreign Capital and Economic Underdevelopment in Jamaica* (Institute of Social and Economic Research, Jamaica).

GRAHAM, RONALD (1982) *The Aluminium Industry and the Third World* (London: Zed Press).

GRAY, EARLE (1982) *The Great Uranium Cartel* (Toronto: McClelland & Stewart).

HOFFMAN, WILLIAM (1974) *Paul Mellon: Portrait of an Oil Baron* (Chicago: Follet Publishing Company).

HOLLOWAY, STEVEN (1980) 'Do Multinational Corporations Benefit from Government Cartel Formation? A study of the Aluminium Multinationals and the Bauxite Cartel' (Ohio State University: unpublished Doctoral Dissertation).

——— (1983) 'Relations Among Core Capitalist States: The Kautsky-Lenin Debate Reconsidered', *Canadian Journal of Political Science*, XVI: 2: 321–33.

INTERNATIONAL BAUXITE ASSOCIATION (1980) *Towards Increased Co-operation in Bauxite Development: Proceedings of the International Conference on Bauxite* (Kingston, Jamaica).

KEOHANE, ROBERT and JOSEPH NYE (1977) *Power and Interdependence: World Politics in Transition* (Boston: Little, Brown).

——— (eds) (1972) *Transnational Relations and World Politics* (Cambridge: Harvard University Press).

KUPER, ADAM (1976) *Changing Jamaica* (London: Routledge & Kegan Paul).

MARVEL, HOWARD (1980) 'Collusion and the Pattern of Rates of Return', *Southern Economic Journal*, vol. 47, no. 2: 236–44.

MIKDASHI, ZUHAYR (1976) *The International Politics of Natural Resources* (New York: Cornell University Press).

MITCHELL, SIR HAROLD (1968) *Contemporary Politics and Economics in the Caribbean* (Athens: Ohio University Press).

MORAN, THEODORE (1974) *Multinational Corporations and the Politics of Dependence* (Princeton: Princeton University Press).

NAVIN, THOMAS (1978) *Copper Mining and Management* (Tucson: University of Arizona).

PECK, MERTON (1961) *Competition in the Aluminium Industry, 1945–1958* (Cambridge: Harvard University Press).

PINDYCK, ROBERT (1978) 'Gains to Producers from the Cartelization of Exhaustible Resources', *Review of Economics and Statistics*, vol. LX, no. 2: 238–51.

PINDYCK, ROBERT (1977) 'Cartel Pricing and the Structure of the World Bauxite Market', *Bell Journal of Economics*, vol. 8, no. 2: 343–60.

REYNOLDS, RICHARD Jr. (1956) *Opportunity in Crisis: The Reynolds Metals Story* (Newcomen Society in North America).

SAMPSON, ANTHONY (1975) *The Seven Sisters* (New York: Viking).

SAMUELSON, PAUL (1973) *Economics* (New York: McGraw-Hill).

STOCKING, GEORGE AND MYRON WATKINS (1946) *Cartels in Action: Studies in International Business Diplomacy* (New York: The Twentieth Century Fund).

STONE, CARL AND AGGREY BROWN, eds (1977) *Essays on Power and Change in Jamaica* (Kingston: Jamaica Publishing House).

TURNER, LOUIS (1978) *Oil Companies in the International System* (London: Allen & Unwin).

UNCEPAL (1979) 'Transnational Corporations in the Bauxite Industry of Caribbean Countries', a paper for the Economic Commission for Latin America.

UNCTC (1981) *Transnational Corporations in the Bauxite/Aluminium Industry* (New York: UN).

VERNON, RAYMOND (1971) *Sovereignty at Bay* (New York: Basic Books).

——— (1977) *Storm Over the Multinationals* (Cambridge: Harvard University Press).

WILKINS, MIRA (1974) *The Maturing of Multinational Enterprise: American Business Abroad from 1914 to 1970*, (Cambridge: Harvard University Press).

WYANT, FRANK (1977) *The United States, OPEC, and Multinational Oil* (Lexington: Lexington Books).

Index